Praise for

"[Decrop] is a highly evolved thinker, and a very strong and thoughtful writer."
—Hal Clifford, editor of David Goggins's *Can't Hurt Me*

"*The Idea Space* demystifies mindfulness for the general public, and presents the topic in a manner that can be accepted by the scientific community."
—Matthew W., *Program Lead at Siemens*

"Decrop possess a remarkable talent for transforming complex subjects into terms that are not only easily understood, but also truly resonate with the reader."
—Dr. Kovalchick, *CEO of PhysioSpace*

"From the vastness of space to the intricacies of the human mind, *The Idea Space* is a transformative exploration that reshapes our understanding of existence and identity."
—ChatGPT, *OpenAI*

THE IDEA SPACE
THE SCIENCE OF AWAKENING YOUR NON-SELF

Clément Decrop

Idea Space Publishing

Idea Space Publishing

The Idea Space™ is a trademark of The Idea Space, LLC.

The Idea Space: The Science of Awakening Your Non-Self, Copyright © 2023
by Clément Decrop. All rights reserved.
No part of this book may be used or reproduced
in any manner whatsoever without written permission
except in the case of brief quotations
embodied in critical articles and reviews.

For information, visit www.TheIdeaSpace.io.

The Idea Space may be purchased for educational, business,
or sales promotional use. For information, please contact us
at the above website.

Cover design and illustrations by Onur Selçuk Akın.
Other design input by Đani Peršić and Tea Lukić.
Edited by Hal Clifford.
Book formatting by Chisom Ezeh.

First printing, 2023.

Library of Congress Cataloging-In-Publication Data In Progress.

ISBN Color Paperback: 979-8-9884102-0-1
ISBN Black & White Paperback: 979-8-9884102-3-2
ISBN Color Hardcover: 979-8-9884102-1-8
ISBN Black & White Hardcover: 979-8-9884102-4-9
ISBN eBook: 979-8-9884102-2-5

To my parents, Sylvie and Georges,
and my brother, Romain.

CONTENTS

FOR THE READER	9
INTRODUCTION	11
CHAPTER 0: THE MAGIC SHOW	21
CHAPTER 1: WELCOME TO THE IDEA SPACE	35
CHAPTER 2: NONDUALITY	53
CHAPTER 3: IMPERMANENCE	69
CHAPTER 4: MINDFULNESS	87
CHAPTER 5: THE ILLUSION OF SELF	109
CHAPTER 6: THE SUNSET CONJECTURE	127
CHAPTER 7: MOVEMENT	151
CHAPTER 8: THE CLOPEN NATURE OF REALITY	177
CHAPTER 9: YOUR IDEA SPACE AS A REFLECTION OF YOUR UNIVERSE	195
CHAPTER 10: IMAGINED REALITIES	233
EPILOGUE: THE RETURN OF THE MAGICIAN	269
ACKNOWLEDGMENTS	283
LEXICON	287
BIBLIOGRAPHY	295
NOTES	301
INDEX	311

FOR THE READER

The following is an excerpt from the introduction of Michel de Montaigne's *Essais*, called "Au Lecteur" ("For the Reader"). It is translated into English and represents the spirit of this book.

C'est ici un livre de bonne foi, lecteur. Je l'ai voué à la commodité particulière de mes parents et amis: à ce que m'ayant perdu ils y puissent retrouver aucuns [touts] traits de mes conditions et humeurs, et que par ce moyen ils nourrissent plus entière et plus vive la connaissance qu'ils ont eue de moi. Je veux qu'on m'y voie en ma façon simple, naturelle et ordinaire, sans contention et artifice: car c'est moi que je peins. Mes défauts s'y liront au vif, et ma forme naïve, autant que la révérence publique me l'a permis. Ainsi, lecteur, je suis (nous sommes) nous-mêmes la matière de mon (ce) livre.

Adieu donc.

De Montaigne, ce premier de mars mille cinq cent quatre-vingt

This book is of good faith. It was passed to a particular community of my family and friends in hopes they recognize me and my humors in it. Why? To enrich *la connaissance* (the knowing) they had of me. This book is meant to be read in a simple, natural, and organic way—without artificial contention. My faults in the book are in my most naïve form or as loud as the public will permit it. Therefore, reader, I am myself the material of this book and I hope you see a reflection of yourself as well.

So farewell.

Decrop, on the tenth of October two-thousand twenty-three

INTRODUCTION

Math and physics are real-world magic. They give humans the ability to describe the world in a specific language, which then allows us to mold our life the way we see fit.

Even with all the scientific discoveries to date, from gravity to DNA to satellites to black holes, it is fascinating to realize the mind itself remains an enigma—even though we use it every day! Although we think we are masters of our domains, rogue ideas continuously bombard our minds to make us reminisce about the inaccessible past or even create fictitious scenarios about the imaginary future. Some of these daydreams are pleasant, some unpleasant. In all cases, they take us away from *reality*, or the present moment.

The Idea Space: The Science of Awakening Your Non-Self explores the scientific foundations of your mind to awaken you to the reality which is hiding in plain sight. As an Ancient once said,

> *"The true thing of things is not hidden—from ancient times till now it's always been obvious."*

Understanding how our mind relates to the universe has been one of life's biggest questions since the dawn of time. Many philosophers and mathematicians such as Michel de Montaigne (1533–1592), Blaise Pascal (1623–1662), Immanuel Kant (1724–1804), and Bertrand Russell (1872–1970) have attempted to answer this mystery in their own way. Many of them contributed their best work at an early age, such as Blaise Pascal at 16 and Albert Einstein at 26.

Today, the problem is no less prominent, and failure to understand the connection between your mind and the universe can lead to a great deal of suffering. Most notably, whenever you attach to a specific thought, idea, or person you are liable to suffer. For instance, if you are in a relationship, you become attached to someone and expect them to behave a certain way.

When they don't meet your expectations, you may feel hurt or angry. This is because you have attached yourself to the idea of how they should behave, rather than accepting them as they are. In this way, attachment can lead to suffering.

This book inspires anyone searching for the hidden truths of our world to find a genuine, sincere, and harmonious purpose to life by allowing you to view your own thoughts, emotions, sensations, and perceptions as objectively as you would view objects in spacetime—devoid of "I." Thereby, *awakening your Non-Self*.

The beauty of these truths is they are hidden in plain sight. All you have to do is *look*.

FRAMEWORK OF THE BOOK

At a high level, The Idea Space revisits the relationship between your mind and the universe with the help of modern techniques. The book provides the beginnings of a digestible structure for the mind that is congruent with modern physics and applicable to all humans. In short, the book represents a union of the humanities, like Zen, and the exact sciences, like physics (figure 1).

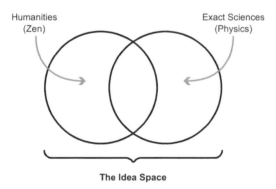

Figure 1. *The Idea Space* is a union of the humanities and exact sciences.

To appropriately study the mind, we turn to science. That said, there are two types of sciences, and understanding their differences is vital to better comprehend how our mind fits within the universe. The *science of objects* is what you are probably most familiar with, and it approximates observations we can measure, like the fields of physics, biology, chemistry, and the other natural sciences (figure 2-a). Conversely, the *science of the first person*, a term introduced by philosopher Douglas Harding, approximates observations that have zero measure, like your *idea space*—a mathematical model of all your thoughts, emotions, sensations, and perceptions (figure 2-b). An object with zero measure simply means it looks like nothing to an outsider.

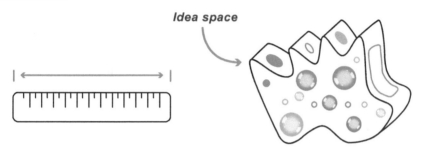

(a) The science of objects deals with things with measure

(b) The science of the first person deals with things with zero measure

Figure 2. (a) Physics, biology, and chemistry deal with the science of objects, while (b) the idea space falls under the science of the first person.

A key tenet found across the science of objects, the science of items we can measure, is "observational proof." In other words, theories are "confirmed" once they line up with observations. However, the true foundation of these sciences lies in falsifiability. Specifically, you can seldom prove an idea to be true. Instead, sciences are an endless cycle of conjectures and refutations, or trial and error. These conjectures show up as theories, while the refutations are decisive observations that defeat the conjecture.

For instance, in elementary statistics, you start with a hypothesis, or assumption. Then, you conduct a test to see how that hypothesis holds up. This is the familiar scientific method. At the end of the test, you can either (a) fail to reject the hypothesis (i.e., not enough information) or (b) reject the hypothesis. At no point do you "confirm" a hypothesis. If you fail to reject a hypothesis through multiple tests, then the hypothesis is said to be "true," or a "natural law."

Thus, the only domain where proofs are allowed, or an idea is said to be true, is mathematics. This is due to the precise nature of the mathematical language. For a trivial example: 2 + 2 = 4. This is true for anyone, at any point in our universe.

The beauty then exists when you take a mathematical concept and apply it to a testable observation. This is the foundation of physics. If the mapping passes testing, then you have a powerful model used to approximate the world. As Galileo wrote, "The great book of Nature is written in mathematical language."

The Idea Space continues this philosophical trend by quantifying a new science for consciousness, the science of the first person, based on mathematical properties and testable observations, to describe the mind. In doing so, the book introduces a new vocabulary for discussing consciousness to promote further exploration into this line of thinking. *The Idea Space* is a scientific and mathematical exploration of consciousness for the layperson that shows you how to view your own thoughts objectively, leading to a happier existence.

While the science of objects has many tools to study its nature, like rulers, sensors, and semiconductors, the main tool we'll use to study consciousness is mindfulness. As meditation teacher Anagarika Munindra wisely notes, "If you want to understand your mind, sit down and observe it."

Therefore, beginning in Chapter 2, whenever we start a new chapter, we'll start by performing a short meditative exercise. Taking the time to practice is key to awakening. As Dōgen Zenji says, "Awakening is abundantly inherent in each individual; nevertheless, without practice it will not be revealed, and without enlightenment it will not be realized."[2] The short exercises allow you to explore your mind from the least unbiased place possible: *beginner's mind*. If you want more meditation exercises, then check out the two complementary meditation card decks, "100 Daily Mediations" and "100 Mindful Prompts," at www.TheIdeaSpace.io.

Additionally, on the website, you'll find supplemental material, bonus chapters, and the Idea Space Whitepaper. The supplemental material serves as a complement to the concepts discussed in the book, while the bonus chapters provide further insights into various aspects of our universe. The whitepaper offers a technical analysis of the concept of an idea space, catering to those interested in a more in-depth exploration.

INTRODUCTION

ABOUT ME

My name is Clément, and I am a human, author, inventor, and entrepreneur. I've led Resilient Leadership events for thousands of participants in over 40 countries on the topics of mindfulness, sleep, exercise, and nutrition. I've also worked with dozens of inventors on Wikipedia's Most Prolific Inventor's List to submit 130-plus patent disclosures in a one-year span, of which 50-plus are filed and 25-plus are issued, as of late 2023.

Overall, *The Idea Space* has been a long time coming. I was born in Brussels, Belgium. Then, at the age of three, I moved to Barcelona, Spain, and at six I moved to Allentown, Pennsylvania.

Arriving in the United States presented a unique set of challenges. Not only was there a massive cultural change, but there was also a completely new language to learn! During my first two weeks of school in the United States, I came home crying every day because I didn't speak English. During this time period, sleep was my greatest fear. "Alone, in the dark, with my own thoughts? No thank you!" I suffered from crippling panic attacks at night that paralyzed my body. Most nights, I couldn't even scream for help because I was so afraid.

To help with my night terrors, my parents introduced me to the wonders of mindfulness by helping me perform simple body scan meditations to help me fall asleep. Thus, from an early age, I developed a keen sense of curiosity as to where thoughts come from and where they go once noticed.

As I grew older, I developed a passion for the sciences. I loved learning how the world worked on a fundamental level. The beauty of math and science is that there is usually a right answer and multiple ways to get it. This foundation helped me build an objective perspective of the world.

To continue my passion for the sciences, I studied Mechanical Engineering at Penn State and worked various jobs in different countries, like France, Spain, the United Arab Emirates, and the United States. After college, I worked at IBM where I became a prolific inventor, led myriad *Resilient Leadership* events across the globe, and drove the sales presentation of a $10 billion cloud deal for the NSA.

Even with all these accomplishments, I was burnt out and fell into a "purpose gap." I could not avoid questions like, What is the meaning of life? What is the bigger picture? How am I adding value to the world? What is my purpose? To answer these questions, I intensified my learnings. During my free time, I continued my pursuit of knowledge by reading graduate-level math and physics textbooks, while simultaneously expanding and deepening my meditative practice. I went on week-long silent retreats to rediscover what it means to be human. I came to understand *there's more to this world than what today's math and physics can explain*. For how great modern physics is, it had yet to provide a sustainable model for consciousness.

The combination of meditation and the exact sciences in my mind was like the merger of two neutron stars: the creation of a supermassive black hole.

What happens if you take the basics of mathematics and apply them to thoughts? Namely, what if there was a space that consisted of thoughts, emotions, sensations, and perceptions—an idea space? What would the properties of an idea space be? What would the *identity* of an idea space be? How would an idea space fit within the world around us? And how could the concept of an idea space benefit humanity?

Thus, *The Idea Space* was born.

All in all, everyone has their own idea space located at the center of their own observable universe—a giant sphere centered on you where everything you see is in the past. Your idea space is unique to you, uncountable, and has zero measure. More on these concepts later in the book. At its heart, an idea space is an exact solution to Einstein's field equation—the core building block of general relativity. Depending on who you ask, there are only six, practical exact solutions used to describe the observable universe and various stellar objects, like stars and black holes. The last of these solutions was introduced in 1963. The book introduces an overview of the seventh solution to describe the mind in a digestible format.

HOW THIS BOOK UNFOLDS

The goal of this book is to awaken your Non-Self. This means developing a truer understanding of what the world is in order to see what the world is not. Most notably, understanding "I," your name and identity, is simply another appearance in your idea space. It is another thought.

This book shows why investigating how we think should not be ignored and how thinking can be looked at through a traditional scientific approach. An idea space is a new tool, or framework, to make meaning of experiences that were once ineffable, like enlightenment. *The Idea Space* dispels the usual mysticism associated with mindfulness and invites the skeptics to investigate their minds for themselves, thereby allowing the reader to develop their own mindfulness practice and collect its benefits, like peace of mind.

To awaken your Non-Self, we'll lift various *veils of illusion*. In other words, we'll introduce many Unknown Unknowns, or ideas you did not know existed, to develop the *Path of Awakening* (figure 3). Not everything discussed here may be an Unknown Unknown to you. If certain concepts are not, then this treatment of them may still be a nice refresher in understanding the bigger picture.

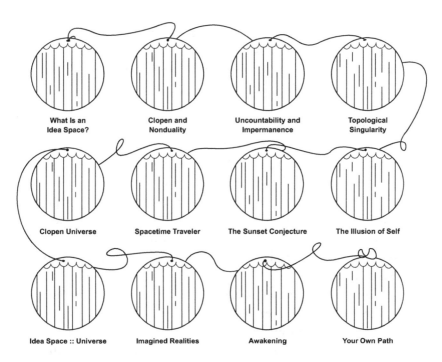

Figure 3. The veils of illusion that build the Path of Awakening.

Each veil is dedicated to a specific chapter. At the end of each chapter, we shall summarize the particular veil lifted, noting the difference in life before the veil was lifted and life after the veil was lifted. By the end of the book, all the veils will be lifted and you will have awakened your Non-Self, thus allowing you to live a more purposeful and harmonious life.

The book begins and ends with a fictional short story to set the mood, tell you everything you need to know about *The Idea Space*, and allow the book to come full circle.

From there, the main content starts by defining what it means to awaken and highlighting some key properties of an idea space. This foundation then leads to a discussion on groupings. Here, we'll uncover the meaning of "clopen," which is a thought-provoking concept synonymous with "nonduality." Afterward, we'll dive into the realm of impermanence through uncountability. "Uncountable" means you cannot count all your thoughts, emotions, sensations, and perceptions. As soon as you count one, another one appears. "Topological Singularity" is a set of objects with uncountable depth and zero measure. Your idea space is an example of a topological singularity. The "Illusion of Self" signifies, as the wise guru Joseph Goldstein says, "The 'I' in mind is extra." After these first four veils, you will have built a robust framework of the idea space, understood mindfulness more rigorously, and glimpsed into your Non-Self.

Next, we'll detail how your idea space relates to the universe at large. To start, we'll see how everyone is at the center of their own observable universe through the *Sunset Conjecture*. Then, we'll see how you are a spacetime traveler—just not in the way you think. From there, we'll see how the universe acts in a clopen manner through "particle-wave duality." Afterward, we'll explore exactly how you can view your idea space as objectively as you would view objects in the universe by understanding how your idea space is analogous to (i.e., "::") to your observable universe.

In the final chapter of the book, we delve into the concept of "imagined realities," or intersubjective realities, which are the collective idea spaces we create. Money, social norms, laws, macropolitical systems, and scientific theories are all examples of these realities. It's important to distinguish subjective ideas, which are dependent on an individual's consciousness and beliefs, from intersubjective ideas, which exist within a communication network linking the subjective consciousness of many individuals. Unlike subjective ideas, intersubjective realities can persist even if one individual stops believing in them. Intersubjective realities can therefore only dissolve or mutate if a significant quantity of people stop believing in them.[3] For example, Facebook is an intersubjective reality that exists in the shared imagination of millions of individuals.

In short, intersubjective realities are only real because enough people believe them to be real. Once you are able to recognize what is an objective reality versus an imagined reality, you can stop wasting time and energy on things that don't matter and focus on things that do, thus reducing much of the stress commonly associated with these imagined realities.

Lastly, we end our Path of Awakening by revisiting what it means to awaken and set you up on your own path. After reading this book, you will experience many more veils of illusion in your journey through life. Be open to the experience. Let the Unknown Unknowns enter your idea space, so you can build a better understanding of what all ~*this*~ is.

The Idea Space: The Science of Awakening Your Non-Self can be read at multiple levels. For instance, you may read this and never consider the content ever again once you put the book down (figure 4-a). This experience is fine, but it is not the one I hope to generate for you. Another level is one where you use imagination as much as possible (figure 4-b). As Albert Einstein said, "The power of our imagination is greater than the power of our intellect." I hope the myriad pictures inspire additional stimulation for you to experience your Non-Self.

(a) Read about your Non-Self

(b) Experience your Non-Self

Figure 4. The different ways to read this book.

The purpose of understanding consciousness as an idea space is it helps disentangle the persistent cravings and suffering tied to ego attachment. Essentially, detaching from "I" is crucial to reduce suffering, as the ego perpetually craves more: "I" want ice cream, "I" want Netflix, "I" want comfort. But what lies beneath this craving? The root of all suffering. The unending desire for more, coupled with our attachment to the objects of our cravings, results in a continuous cycle of dissatisfaction. No matter how much we attain, it only leads to further craving. It is akin to attempting to fill a bucket with a hole in it—no matter the volume of water poured, the bucket will never be full. As guru Naval Ravikant astutely observes, "Desire is a contract to be unhappy until you get what you want."

To overcome this cycle of craving and suffering, we must learn to detach ourselves from our desires. We must learn to let go of the constant need for more and find contentment in what we have. This is not to say that we should stop pursuing our goals or stop enjoying the things we love. But we should approach them with a sense of detachment, without the attachment that leads to suffering. A simple way to experience this is by periodically imagining you are living someone else's dream life. This will help develop perspective for the present moment and build a sense of gratitude.

By viewing our thoughts as an idea space, we can not only divorce ourselves from "I" and the constant craving that comes with it, but also find peace and happiness in the present moment. Vietnamese monk Thich Nhat Hanh states the truth around happiness wonderfully, "Happiness is available. Please help yourselves to it."

Chapter 0
THE MAGIC SHOW

"All sayings and writings return to one's self."
- Pai Chang

Imagine yourself living in the year 2,432 on Mars (figure 5). It has been 321 years since Mars has been inhabited by humans, and it is now almost completely terraformed: trees grow abundantly, rivers create luscious landscapes, and the once-active volcano Mount Olympus Mons is now a popular ski resort. Mountains and rivers make Mars the most beautiful place in the solar system.

Figure 5. Mars circa 2,432.

Even with all these changes, humans still follow various social norms reminiscent of the year 2,000: *quantum crypto* is the currency of use, *Arurrak* is the main language, the *Galactic Trade Federation* is the new body of governance, the *Rights of Sapiens* is the law of the land, and skyscrapers dominate the Martian skyline (figure 6).

Figure 6. Skyscrapers dominate the Martian skyline.

One evening, you decide to go to a magic show called *The Idea Space*. Clouds cover the gray skies, while rain and fog fill the streets. You hop on your hover bike and fly over to the show. The narrow streets are packed, as thousands of busy Martians are flying to their destinations. You whiz through traffic and make your final turn to see the entrance of the theatre behind a small alley. *Huh—I've never seen this place before.* You hop off your bike and lock it up.

Thunder rumbles in the distance. You make your way down the alley, walk down a flight of stairs, and open a heavy, metal door. On the other side lies a dark, narrow hallway with a red door at the end. You pull back your hood and slowly walk across the hall, looking for anything out of the ordinary. You push open the doors to enter a small room with a person sitting behind a call window. The light behind him flickers.

"One ticket for *The Idea Space,* please," you say in Arurrak.

The person behind the glass looks at you and says, "That'll be 25 quantum." You nod your head as he scans your eyes to transfer the funds for a digital copy of the ticket. The man hands you a pamphlet and says, "Enjoy the show."

You walk past the call window to a set of double doors. On the other side lies a dimly lit theatre with about 50 seats. As the usher scans your eye and takes you to your seat, you note the theatre is about half-filled. When you sit down, the usher says, in an eerily similar voice to the person at the call window, "Enjoy the show." *Huh—that was odd . . .*

In your seat, you read the pamphlet to see what the evening has in store:

WARNING:
DO NOT ATTEND THIS PLAY UNLESS YOU WANT
TO CHANGE THE WAY YOU THINK.

Welcome to The Idea Space! Before starting the show, I ask the attendee to come with beginner's mind. If he or she does not know how to do this—no worries—this can always be accomplished by focusing on the breath. Therefore, at any point before the show, please do a full S.T.O.P.P. For convenience, the steps are listed below. If you prefer to take a couple of breaths with your eyes closed, then that works too. Simply get situated and make yourself comfortable.

STOP. *Stop what you're doing. Stop thinking. Stop getting distracted. Just stop.*

TAKE A BREATH. *Take three deep breaths. Breathe so there is no pause between breaths.*

OBSERVE. *What do you notice? Your breath? A thought? A sensation? Simply notice.*

PURPOSE. *What do you want to get out of this show? Learn something new? Up to you.*

PROCEED. *Now that you've achieved beginner's mind, we can proceed.*

As the show goes on, please continue focusing on your breath. You may forget at times, and that is totally acceptable—forgetting is what makes us human. If at any point in time you feel you need a breather, then you are more than welcome to take one. The show will wait for you.

You pause, gently close your eyes, and take a couple of deep breaths. As you sit in your chair, you feel gravity pulling you down. You sense the air traverse through your nostrils to the back of your throat and down toward your lungs. With every in breath, you feel your body expand with fresh energy. With every out breath, your mind and body are immediately more relaxed. You are now in a state of pure and utter equanimity.

As you continue to focus on your breath, the light gradually dims until the room is almost completely dark. The murmurs of the crowd fade into silence. The show is about to begin. Suddenly, a bright, circular spotlight illuminates the stage, revealing a large, red veil. The audience holds its breath in anticipation. With a dramatic flourish, the veil is lifted, revealing a solitary figure standing on the stage. It's a magician, gazing out into the darkened space that is his audience.

The magician wears a classic magician's outfit: black shoes, black pants with white stripes up the sides, a black vest with a boisterous blouse underneath, a sleek black jacket on top with a white handkerchief tucked into the square pocket, a black cape with red lining on the inside, and a top hat. The magician is looking down with his hand on top of his hat so you can't see his face. The theatre is dead silent.

After a few seconds, he looks up and shouts, "Welcome to The Idea Space!" *Huh—isn't that the same voice as the usher and the person at the call window?* The magician starts pacing back and forth on stage, saying, "This is a play in three acts. First, we'll demonstrate what an idea space is. Second, you will explore your idea space for yourself. And, if you're lucky, you'll experience the third act, which will reveal the main trick. In fact . . . The main trick has already started! It's been staring you in the face since the moment you decided to attend this show . . . But! I cannot reveal it now! We must do a couple of magic tricks to warm you up first."

The magician stops, looks at the crowd, and says, "In order to perform my first act, I will need a volunteer. Do I have any takers?"

As he says this, you see a couple of people raise their hands. You stay silent in your seat *hoping* you don't get picked. Then, mistake, you lock eyes with the magician. He points at you, "You—there." *Oh god, not me.* The magician points again with his other hand—both index fingers now pointed at you. "Yes. You. Why don't you come up here to help me with a trick? This is a magic show, after all."

As soon as he says this, you hear clapping among the crowd. *Oh god, why did he have to choose me?* You blink, and before you know it, your seat magically transports to the stage and has taken on a different form, too. It is now a throne covered in red velvet with a golden frame (figure 7). You gulp—out of fear or excitement, you do not know.

THE MAGIC SHOW 25

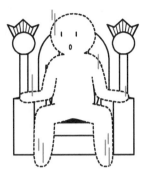

Figure 7. Your new seat.

As the magician stands behind you with both hands on the rounded tips of the throne, he states, "An idea space, or the human mind, is a mystical creation of infinite depth, hidden from the outside world . . ."

The magician approaches you and directs his attention to the crowd. "Can anyone see all the thoughts, emotions, sensations, and perceptions of our dear volunteer?" he asks. "Can anyone here see what appears in the consciousness of our dear volunteer?"

As he says this, you hear murmurs from the crowd: "No," "I don't see anything," "There's nothing there!"

"Precisely!" says the magician. "Our volunteer's idea space, which consists of thoughts, emotions, sensations, perceptions, and the empty set, looks like nothing! However, simply because it *looks* like there is nothing there doesn't mean there is nothing there. We can make an object smaller and smaller until it has zero measure—at which point, it looks like nothing!"

To demonstrate, the magician pulls off his hat and shows you the inside of it. He asks, "For instance, dear volunteer, is there anything in this hat?" You shake your head no—*there's nothing there*. The magician then shows the inside of the hat to the rest of the audience and says, "Clearly, it *looks* like there is nothing in this hat . . . But! What if I told you behind the veil of nothingness lies an animal so small, it looks like nothing?" You hear the crowd murmuring.

"Dear volunteer, please hold this hat for me," the magician demands as he hands you his hat. You firmly hold the hat with two hands at the brim. The magician pulls out a wand from his sleeve and says, "For my first trick,

I will make the animal hidden in this hat larger and larger, until it takes the shape of an animal we rarely get to see here on Mars."

The magician waves his body and wand around as if he is doing a weird dance, points the wand at the hat, and exclaims, "ABRACADABRA!"

You feel the hat get heavier and heavier. *What is happening?* Two small ears pop out of the hat. Soon enough, a whole rabbit appears (figure 8). As this happens, you hear audible gasps from the crowd. The magician takes the hat back from you and pulls out the rabbit. "Aha!" he shouts. "What once looked like nothing shows its true colors: a bunny!"

Figure 8. Out of nothing comes a rabbit.

"The catch: there was never nothing there! The bunny was simply so small, it looked like nothing!" The magician puts the bunny in your hands, and you gently pet the creature. *So this is what a rabbit looks like!*

He starts pacing back and forth once more. "To make this clear, as our dear volunteer holds the rabbit, we can return it to its original form by simply performing the opposite spell!" As the magician says this, he does what seems to be the opposite of his weird dance from before, then exults, "ARBADACARBA!"

You feel the bunny in your hands get lighter and lighter. *It is shrinking!* Eventually, it becomes so small it disappears (figure 9)! You hear people standing in the crowd trying to get a closer look. After a couple of seconds, all that's left is—nothing!

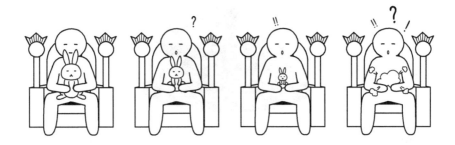

Figure 9. You can make a bunny so small, it eventually looks like nothing.

The magician laughs seeing the stunned faces in the crowd, "Hahahah! Do not worry, friends, our bunny is completely safe and in a special place. Although it looks like there is nothing in our volunteer's hands, there is in fact a bunny of zero measure and infinite depth. Since it has zero measure, it looks like nothing . . . "

The audience members sit back down flabbergasted as to what they just saw. You wipe the bead of sweat from your forehead as the spotlight's warm light shines bright on you. The magician stops pacing and says, "See, our volunteer's idea space is just like this bunny. Although it looks like there is nothing there, behind the veil of nothingness lies an uncountable depth filled with thoughts, emotions, sensations, and perceptions."

The magician approaches you, finger-tenting his hands in a dastardly way, like an evil villain who just hatched a genius plan. He states, "You see, I have come across a marvelous invention from the great encyclopedia, *A Hitchhiker's Guide to the Galaxy*. It is an evolution of the old supercomputer *Deep Thought* called *Deep Mind* . . . Its purpose? To show the idea space of any living creature."

The crowd murmurs with excitement having heard legendary tales of the Deep Thought computing system: "Deep Thought—I thought that was a myth?," "From what I've heard, Deep Thought said the meaning of life is 42!," "What is Deep Thought?"

Out of thin air, the magician conjures a rickety contraption (figure 10). "This, ladies and gentlemen, is Deep Mind. To see the magic that is, we'll put it on our dear volunteer's head to see their idea space."

Figure 10. *Deep Mind* is able to show the idea space of any living creature.

The magician wastes no time putting the apparatus on your head. After a few seconds of making sure everything is in place, he presses the ON button. Sparks crackle as Deep Mind starts to make a loud buzzing noise.

"You see, ladies and gentlemen," shouts the magician, "within a few seconds of turning on this machine, the real world magic happens!" As he says this, Deep Mind shoots a beam out of the device and a hologram of your idea space reveals itself to the audience (figure 11). The electricity in the air is deafening.

Figure 11. A hologram of your idea space shoots out Deep Mind.

THE MAGIC SHOW

Sparks fly toward the audience as Deep Mind seems to generate a great deal of energy. The magician points to the hologram and says, "Again, what was once hidden behind the veil of nothingness reveals its true colors: our volunteer's idea space! Everything our volunteer can sense, think, perceive, and feel lies within this beast."

As he says this, your idea space quickly takes on a new form (figure 12). Even more sparks fly out of Deep Mind as a thunderous crack whips through the air. The crowd looks bewildered.

Figure 12. Your idea space is in a constant state of flux.

The magician laughs, "Hahahahah! Worry not, friends—Deep Mind is harmless!" As he says this, your idea space continues to wobble, unable to stay still.

The magician continues, "Ahhh . . . the good old *idea space wobble*. The reason an idea space wobbles is because it is *uncountable*, or impermanent. It is in a constant state of flux. You cannot count every thought, emotion, sensation, perception, or emotion present. As soon as you count one idea, another one appears!"

You look at the magician in a mild state of panic. The magician looks back at you and gives you a gentle smile. For some reason, the fear within you dissipates. The magician looks back at the crowd and says, "Dear guests, if you look underneath your seats, you'll find each of you now has your own Deep Mind machine. Please put them on your heads and turn them ON. The switch is right there in the back."

The audience members all grab their respective Deep Minds from under their seats. One by one, they put Deep Mind on their heads and turn it on. Sparks fly, lighting up the room as the idea space of every guest now becomes visible to the world (figure 13).

The magician smiles and says, "You can now see every thought, perception, sensation, and feeling of every other person in this room. These are all your *personal idea spaces*."

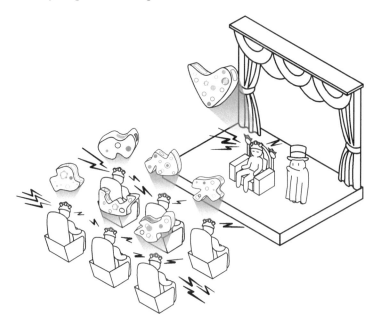

Figure 13. Everyone has their personal idea space showing.

The magician lifts up his wand and states, "The real magic happens when personal idea spaces mix, thus creating a *macro idea space*! This can be through verbal or even nonverbal interactions." He waves his wand, and everyone's Deep Mind starts shooting sparks as a new hologram takes shape (figure 14). Electricity fills the room as loud crashes fill the air.

Figure 14. An idea space among a group of people is a macro idea space.

As sparks fly in the air, the magician says, "Voila! The macro idea space also looks like nothing and is uncountably deep, or impermanent. It is constantly changing to create the imagined realities of our world, like quantum crypto, Arurrak, the Galactic Trade Federation, and even the Rights of Sapiens."

The magician claps his hands twice, and in an instant the sparks are gone. No more hologram. The room is dead silent. You touch your head, but don't feel anything there. The crowd's Deep Minds have vanished too. The magician seems to be fiddling with a coin—effortlessly rolling it back and forth across his fingers.

"Hehehe," he chuckles, "you see, idea spaces can be quite powerful. But! If we are not aware of our idea space, then it can control, or even domesticate us! How many times a day do you get angry at a situation that did not even happen? In those instances, can you notice you are simply mad at thoughts?! At something or someone nonexistent!?"

You hear confusion in the crowd. Sweat keeps rolling down your forehead from the bright spotlight. As you wipe the beads of sweat, you look at the magician, who appears to be inspecting the coin.

"Now," says the magician, "Act II is focused on you discovering your idea space for yourself. But first! A sneak preview of the main trick . . ." He pauses and looks up. "A simple question: *Did gravity exist before Isaac Newton brought forth the idea of gravity?*"

As the magician scans the crowd—DING!—you hear him flip the coin in the air. Time slows down. You look at the faces in the crowd: all their mouths are gaping (figure 15).

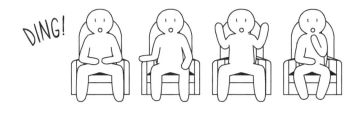

Figure 15. The stunned faces in the crowd.

The crowd gasps as you hear—ding, ding, ding—the coin hit the ground. You turn—the magician is gone. He is nowhere to be seen.

As the crowd gets out of their seats to see what has happened, you also stand to inspect the stage. There is no trace of the magician! A tiny light glints on stage—the coin the magician was holding. You kneel to pick it up. One side reads: "Liberty, In God We Trust, 2026." The other side reads: "United States of America, Quarter Dollar." *Huh—this must be old. The dollar and United States haven't been around for at least 200 years.*

The lights slowly rise as the auditorium begins to empty out. You put the coin in your pocket and follow the others. The will-call window is empty, and there is no sign of the usher. As you open the heavy metal door to walk outside, you see the rain has stopped. The sun is setting and Mars's two, tiny moons rise in the sky. You hop on your hover bike and make your way back home. *Huh—I wonder what the main trick was . . . We didn't even get the chance to explore our own idea space* . . .

When you arrive home you find a package addressed in your name (figure 16). All it says is: *Act II*. You rush inside and open it to find a book. Eager to learn more, you sit in your favorite chair and begin reading . . .

Figure 16. A package in your name that says: Act II.

Chapter 1
WELCOME TO THE IDEA SPACE

"If a picture is worth one-thousand words,
then an idea is worth one-thousand pictures."

Awakening is to know what reality is not.[4] For instance, which inner circle *looks* bigger (figure 17)?

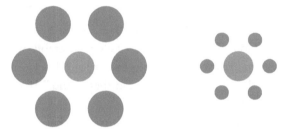

Figure 17. Which inner circle looks bigger?

Clearly, the one on the right does. The catch, which you may already have guessed, is both circles are the same size. Here, we get a glimpse into awakening. You see one thing; reality is another. The moment lasts only an instant, but its effects linger for an eternity.

Let's look at another example. Figure 18 has two horizontal lines with fins pointing in opposite directions. Which horizontal line *looks* the longest?

Figure 18. Which horizontal line looks the longest?

The bottom line does. No matter how you look at it, the bottom line will always look longer. Now, if you get a ruler to measure the two, you will see they are the same length. This is the famous Müller-Lyer Illusion. Once again, we see one thing, but reality is another.

Good things come in threes. For the last exercise, try to answer the following as intuitively as possible. Do not linger on the riddle for too long. Instead, state the first answer that comes to mind.

A bat and a ball cost $1.10.
The bat costs $1 more than the ball.
How much does the ball cost?[5]

Think. Think. Think. Quickly come to an answer. What is it? Congratulations! If you are like most people on this fine Earth, you probably guessed ten cents. And, if you are like most people on this fine Earth, then that is the wrong answer. What is ten cents, plus one dollar and ten cents? One dollar and twenty cents. So, that's not it. The right answer is five cents. Honestly, getting this wrong is nothing to be ashamed of! Most people, including yours truly, also failed miserably when presented this problem for the first time. As philosopher John Locke said, "Men give themselves up to the first interpretation in their mind." If you got it right, then I hope you still get the message.

The point of these exercises is to illustrate what awakening is. It is the moment when we realize our perceived notion of reality may not be what reality actually is. In all cases, there is an instant of *enlightenment*. An "aha!" moment. The sensation you feel during this enlightenment is pretty ineffable. The closest experience that captures the sentiment is the feeling you get after a good magic trick. It is as if you became aware that a veil was lifted.

This book is threaded with different veils like these to awaken your *Non-Self*. Your Non-Self is the simple fact that "I," your identity, is merely another thought in your *idea space*. In reality, your name is an idealization others use to approximate who you are. It is only one layer of your true Self. You are so much more than your name. You are more than your thoughts. You are more than your emotions. Your Non-Self consists of a multitude of fractal layers including your molecules, your cells, your name, your country, your planet, your galaxy, and your universe. After all, everyone lives at the center of their

WELCOME TO THE IDEA SPACE

own observable universe . . .

In this chapter, we'll take a look at the basic properties of your idea space, before dissecting them more thoroughly in the following chapters.

BASIC PROPERTIES OF AN IDEA SPACE

Your idea space consists of your thoughts, emotions, sensations, perceptions, and the empty set, ∅, or nothingness. Your idea space is unique to you, uncountably deep, has zero measure, and is located at the center of your observable universe.

"Uncountable" means it is impossible to count all your thoughts, emotions, sensations, and perceptions. In other words, your idea space is *impermanent*, or in a constant state of flux. Put plainly, notice an idea and another idea appears.

"Zero measure" means your idea space looks like nothing to an outside observer. More specifically, in normal geometry, the distance between two points, *a* and *b*, is simply: *b* - *a*. So, if an object has zero measure, then the two points *a* and *b* are so close together they look like the exact same point. At which point, the distance between the two looks like nothing.

For instance, imagine we have a line (figure 19-a). Clearly, there is a distance between the two sides of the line. Now, let's make the line smaller (figure 19-b). And even smaller (figure 19-c). And, eventually, so small the line has zero measure. Essentially, there is no distance between the points on the line, but it is still a line. When the line has zero measure, it looks like nothing (figure 19-d).[*]

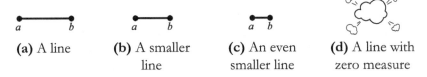

(a) A line (b) A smaller line (c) An even smaller line (d) A line with zero measure

Figure 19. A line with zero measure looks like nothing, but there is still a line there.

---

[*] Formally, a zero set is a set such that the distance between a and b is smaller than ε, for all $\varepsilon > 0$. In Lebesgue Theory, the outer measure of a zero set is zero (m*0 = 0), and the outer measure of the empty set, ∅, is also zero (m*∅ = 0). So, in a sense, zero and nothing look the same as they both have an outer measure of zero.

Now, zero, 0, is not equal to the empty set, or nothingness (i.e., ∅ ≠ 0)! The conundrum is the two look the same (figure 20). For instance, say I took a ruler and measured something with zero measure. Obviously, I would get zero, 0. Now, if I took the same ruler and measured nothing, ∅, then I would also get zero, 0! In other words, something that has zero measure looks empty; and, something that is empty looks like zero. However, simply because it *looks* like there is nothing there, doesn't mean there actually is nothing there. As Martine Rees said, "Absence of evidence is not evidence of absence." In reality, something with zero measure can be uncountably deep, like your idea space.

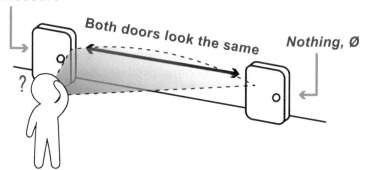

Figure 20. An object with zero measure looks like nothing to an outside observer.

To locate your idea space, simply perform the following experiment: hold something in your hand, like a pencil, a book, or a computer. Clearly, you can measure it. You can feel it. Others can see it. Now, close your eyes and bring a mental image of what you're holding into your mind. Can anyone else see the image in your head? No—it has zero measure and therefore looks like nothing to an outside observer. But is there nothing there? No—clearly there is something there. Something only *you* can see. This holds true for all your thoughts, emotions, sensations, and perceptions. No one else can see them but you. *That* is your idea space.

As we shall see later on the book, your idea space sits at the center of your own observable universe, which is a giant sphere centered on you where everything you see is in the past.

WELCOME TO THE IDEA SPACE

ELEMENTARY IDEAS

The best way to understand an idea space is to give it properties to develop ubiquitous structure. In other words, we must develop a shared language, so we can effectively communicate and ensure that we are discussing the same concept, thereby creating a common layer of understanding.

We start by considering the types of elements in your idea space. Let's creatively call these elements *ideas*. For simplicity, let's assume an idea can be represented as a circle (figure 21).

Figure 21. An idea is the building block of our idea space.

There are many different types of ideas. Drawing inspiration from the five aggregates of Buddhist philosophy, the main *elementary ideas* are thoughts, emotions, sensations, and perceptions. Thoughts consists of words, pictures, memories, daydreams, songs, etc. Emotions involve feelings of pleasant, unpleasant, neutral, and everything in between. Sensations include the classic five: touch, sight, sound, taste, and smell. Perception is one's ability to recognize something. For instance, a pen is a pen. A computer is a computer. Together with the empty set, ∅, or nothing, these elements make up your idea space (figure 22).

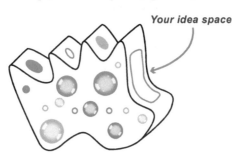

Figure 22. Your idea space consists of thoughts, emotions, sensations, perceptions, and nothing.

You may have counted four elementary ideas, yet five aggregates. The fifth aggregate, which arises from the other four, is *consciousness*. In other words, consciousness is the ability to apply awareness to one or more of these elementary ideas. It is like putting a spotlight onto a particular portion of an idea space. Consciousness is a tool to cognize an idea space.

NESTING IDEAS

Nesting means putting one group underneath another group. For instance, I can have an idea space around the topic of physics; and, within that idea space, I can have subideas of black holes, quantum mechanics, atoms, gravity, etc. (figure 23). Each subidea is an idea space in its own right that lives in the larger idea space.

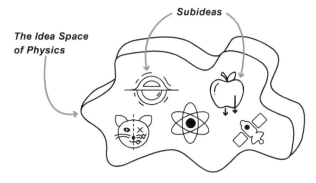

Figure 23. The idea space of physics is nested with its subideas.

EMPTINESS

The concept of emptiness is explored in many meditation practices, so it is important to define it precisely. Emptiness is the opposite of everything. It is nothing. It represents the absence of any elements, and it is denoted by the empty set symbol, ∅. A set is a group of things. So, the empty set is a group of nothing!

The empty set is a trivial yet important finding. An easy way to understand it is as follows: Imagine you have three apples. Each apple has distinct characteristics that can be used to describe it. Table 1 is as easy as it gets.

Table 1. Apple characteristics.

Trait	Apple 1	Apple 2	Apple 3
Color	Red	Green	
Weight	80 grams	100 grams	70 grams
Stem? (Y/N)		N	Y

Clearly, this table is missing entries. That's the empty set, ∅! Your mind may quickly try to fill in those gaps by thinking something like, *the third apple is yellow!* However, you and I will never know what that information is because it is missing. There is simply nothing there.

Now extrapolate the same concept of the empty set, ∅, to the idea space. The empty idea space is often dubbed *beginner's mind*. It is simply a place where no ideas exist, or the space between breaths. In beginner's mind, there are no preconceptions or judgments. You approach each situation with an open mind, free of the influence of past experiences and knowledge. This allows you to see things *yathabhutam*—as they truly are, without any distortions or biases.

The empty idea space is a beautiful concept because it is the best mindset to undertake when creating or exploring new idea spaces. Hence why we'll start each subsequent chapter with a short, meditative exercise. When you start with the empty space, or nothing, you have no other ideas in your mind that can muddle the waters. Ideas can combine in weird and wacky ways we do not understand. So, having the power to start with nothing, ∅, prevents old ideas from negatively mingling with new ones in and is a powerful tool all humans possess.

Beginner's mind can be achieved through a simple mindful pause technique called *S.T.O.P.P.* As you do this short exercise, try to notice what happens in your mind as you go through it. Afterwards, we'll debrief the experience together.

> **STOP.** *Stop what you're doing. Stop thinking. Stop getting distracted. Just stop.*
> **TAKE A BREATH.** *Take three deep breaths. Breathe so there is no pause between breaths.*

OBSERVE. *What do you notice? Your breath? A thought? A sensation? Simply notice.*
PURPOSE. *Now, what do you want to get from this book? Up to you.*
PROCEED. *And, now that you've achieved beginner's mind, we can proceed.*

When you S.T.O.P.P., thoughts no longer linger in the mind. In figure 24-a, we have a basic idea space broken up by elementary ideas. Step One: As soon as you stop, your idea space vanishes (figure 24-b). Step Two: When you focus on the breath, the only appearances in consciousness are sensations (figure 24-c). Step Three: The mind is a fickle beast. A thought enters the mind, in turn triggering an emotion (figure 24-d). Step Four: Set your intentions to choose which thoughts and emotions to focus on and which to let go (figure 24-e). Step Five: As you proceed, a new idea space is formed with everything you want inside of it and everything you do not want outside of it. Note how perceptions do not form quite yet! Thus is the beauty of beginner's mind—you take everything as is.

It is worth reiterating the fact that emptiness, ∅, is not the same as zero, 0. Although the two look the same, behind the veil of nothingness can lie an object with zero measure and infinite depth, like your idea space.

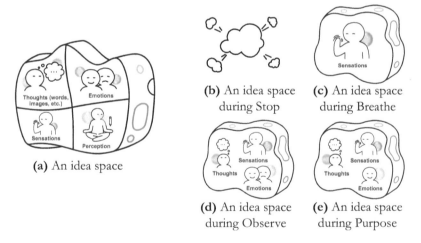

Figure 24. Achieving beginner's mind by performing a mindful pause. **(a)** An idea space before a S.T.O.P.P. **(b)-(e)** An idea space evolving throughout a S.T.O.P.P.

FUNDAMENTAL IDEAS

There are many ways to slice and dice an idea space. We have already done so with our four types of elementary ideas: thoughts, emotions, sensations, and perceptions. Then, consciousness acts as a spotlight on our idea space.

Another way to slice and dice an idea space is through *fundamental ideas*: Known Knowns, Unknown Knowns, Known Unknowns, and Unknown Unknowns. You will quickly note that fundamental and elementary ideas coexist. For instance, you can have a thought you know (a Known Known) or an emotion you have never had and do not even know exists (an Unknown Unknown). They are simply different ways of breaking down an idea space.

At a high level, fundamental ideas dictate what information you know relative to the world and information the world knows relative to you. In the *Known-Unknown Matrix*, the first word is what you know; the second word is what the world knows (figure 25). We'll break down each type of idea in the coming paragraphs, but one thing to note beforehand is these four fundamental ideas are unique to the individual. What one person knows or does not know is going to vary from person to person.

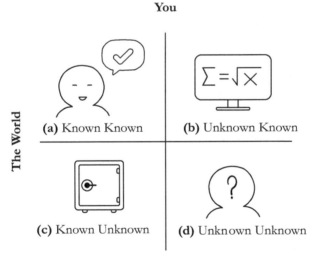

Figure 25. The *Known-Unknown Matrix*. **(a)** Ideas you know to be true relative to the world. **(b)** Ideas the world knows, but seem like variables to you (i.e., you "kinda" know something). **(c)** Ideas that you know, but the world doesn't. **(d)** Ideas that no one in the world knows.

Known Knowns. These are ideas you know to be true relative to the world. A few examples: apples are red (sorry if you are colorblind); the Earth is a close approximation to a sphere (yes, I know there is a bulge at the equator); no number from 1 to 999 includes the letter "a" in its word form; you have seven days left to live after reading this—just kidding. You get the point.

Unknown Knowns. This is information people have discovered, but you have yet to discover for yourself. In other words, something you do not fully know, but you know exists. A lot of times, these are represented by variables, like *x*.

For example, what is the gravitational acceleration, *g*, of an object on Earth's surface?[*] Unless you are a physicist, engineer, or interested in the particular topic, then you probably have no idea. Well, someone smart figured it out and deduced that the gravitational constant on Earth is approximately 9.81 m/s^2. This means when you jump out of a plane, you will accelerate toward Earth at 9.81 m/s^2. In your mind, you may think of this as *gravity*. Each planet has its own gravitational constant. For instance, Mars has a gravitational constant of 3.71 m/s^2. Jupiter: 24.8 m/s^2. The Sun: 274 m/s^2. Can you imagine how it would feel accelerating that fast toward the sun?

Known Unknowns. A Known Unknown is something that exists, but only you know. For instance, there are a lot of different "yous" reading this right now. I bet each of you knows something everybody else reading this doesn't know. For instance, what color is your underwear? I hope you, and only you, know this!

Furthermore, "you" does not have to apply only to a person, for a person is not the only thing that holds information. A good example of this is *black holes*. Black holes have this mysterious veil, or *event horizon*, that prevents anything that falls in from going out of it. In a way, information is trapped inside the surface and is hidden from the outside world—just like the color of your underwear is (hopefully) hidden from the outside world.

Again, Known Unknowns are things you know, but no one else knows.

Unknown Unknowns. This one is my favorite. Things you do not even know you do not know. The interesting part about these is they still play a vital role in dictating the way we live our lives.

---

[*] Acceleration is measured in the units of m/s^2, or meters per seconds squared.

Let's look at one of our first examples: gravity. Before the introduction of the "idea of gravity", it would have been unthinkable to comprehend such a radical and preposterous idea. And yet, today, this concept is crucial in solving some of the world's most mechanical problems, from launching spacecrafts to making our cars run. This leads to an ominous question: *Did gravity exist before Isaac Newton presented it?* It was an Unknown Unknown.

Another way to exemplify Unknown Unknowns is with the "idea of an idea space". Prior to this book, the term "idea space" never really existed and meant nothing. So, before introducing this term, did an idea space exist? It, too, was an Unknown Unknown.

VEILS OF ILLUSION

The big question you can draw from these fundamental ideas is: *Do Unknown Unknowns sit within your idea space or outside of it?* Simply put, Unknown Unknowns must exist outside your idea space, since you cannot hold knowledge of something you do not even know you do not know! Therefore, every time an Unknown Unknown enters an idea space, *a veil of illusion* is lifted (figure 26), providing a crucial step toward achieving an awakened, *buddha* mind.

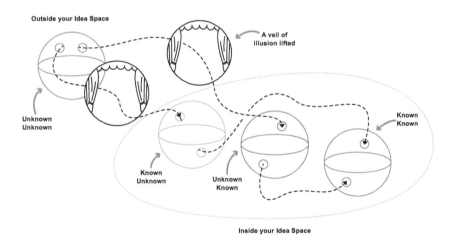

Figure 26. A veil of illusion is lifted once an Unknown Unknown enters an unsuspecting idea space.

In Sanskrit, "buddha" means "awakened one," and it is an enlightened state of being that arises after multiple strings of enlightenment, or awakenings. The catch is a buddha mind does not awaken into anything special. As an Ancient once said, "Absence of attainment is true wisdom; if there is something which is attained, this is just semblance wisdom."[6] In other words, a buddha mind simply awakens to what is already present: sights, sounds, thoughts, emotions, perceptions. As such, a buddha is an ordinary man or woman. As the Zen philosopher Alan Watts says, "A buddha is a man of no rank. He is not above, like an angel; nor below, like a demon."[7]

The sensation of lifting the veils of illusion can be likened to the exercises at the beginning of this book when you realized your perception did not match reality. For example, the one horizontal line looked longer than the other, but it turned out both lines were the same length. Therefore, a veil of illusion being lifted is a moment of awakening—you develop a truer understanding of what the world is to realize what the world is not. When enough veils of illusion are lifted, you can awaken your Non-Self and develop a truer understanding of the world.

The purpose of this book, then, is not only to understand the concept of idea spaces and the significance of lifting the veil of illusion, but also to inspire and guide you on your journey toward achieving an awakened mind—the ultimate goal of a buddha. This is why the book is threaded with the veils of illusion.

PERSONAL AND MACRO IDEA SPACES

There are two types of idea spaces: *personal* and *macro* idea spaces. A personal idea space is the one you are living right now (figure 27-a). Your own thoughts, emotions, sensations, and perceptions. Macro idea spaces occur when personal idea spaces mix (figure 27-b). This can be in the form of written, verbal, or nonverbal communication—for example, you talking to your friends, a town hall, or even reading a book. It's important to note both personal and macro idea spaces retain the general properties of an idea space (i.e., zero measure and uncountable depth). In other words, they both look like nothing and are constantly changing.

WELCOME TO THE IDEA SPACE

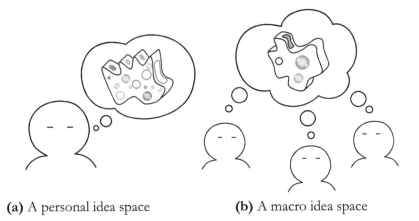

(a) A personal idea space   (b) A macro idea space

Figure 27. Personal vs. macro idea spaces.

The mixing of ideas can occur both in your own idea space or between the idea spaces of various people. When mixing occurs, a lot of things can happen. For example, ideas can meld together to create a novel idea space. This is usually called *inspiration*, and inspiration is the purest form of ecstasy. Other times, ideas can meet and nothing can come out of it. The ideas simply do not mix, and you get a sort of—blah. For instance, two people from different cultures may have trouble communicating effectively because their personal idea spaces are so different.

Overall, the combination of ideas can simply be summed up by the terms *union*, *intersection*, and *difference* (figure 28). A union is the complete combination of all parts of the specific ideas. The intersection is combining only what is in common between ideas. Lastly, the difference is combining only what is unique between each idea.

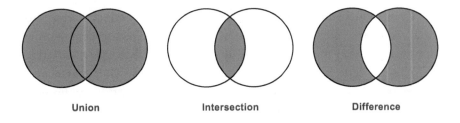

Figure 28. What can happen when ideas meet. Union (left), Intersection (middle), Difference (right).

To illustrate the distinction between all three, imagine you were going on a road trip with your friend. You both have different ideas of what to do and see along the way. You want to go to the zoo and on a hike. Your friend wants to go to the zoo and swim. The union would be to create a plan that includes all the places and activities you both suggested: zoo, hike, and swim. The intersection would be to create an itinerary based solely on the activities you both have in common: zoo only. The difference would be a trip based on suggestions only one person made: hike and swim.

Of course, ideas are so much more complicated than mere circles. Furthermore, there is the possibility of partial unions, partial intersections, and partial differences, which means only bits and pieces of ideas will be considered. That said, these basic properties of union, intersection, and difference are sufficient for our purposes.

TRANSFERRING IDEAS

The transfer of ideas is seldom perfect. For example, think of an apple. A nice, juicy, red apple. Picture that shiny apple sitting crisply on a patch of grass with blades curling around it in the wind. The apple's stem is short, but not too short. In the field of grass, there are some flowers next to the apple that contrast nicely with the apple's glare.

What comes to mind when you read these words? Maybe you see an image similar to what I am seeing, maybe not. Even if you think your apple and my apple are close, the apple in my idea space will be minutely different from the apple in your idea space. Words do a poor job transmitting ideas, which is why this book is full of pictures. *If a picture is worth a thousand words, then an idea is worth a thousand pictures.*

Even though the transfer of information is not perfect, it does not mean we cannot convey a message of what an apple is to one another. However, your red apple and my red apple are, and always will be, different.

This concept holds true for other sorts of ideas. For instance, in evolutionary biologist Richard Dawkins's book, *The Selfish Gene*, he gives a great, biological example of birds transferring ideas. This example, first introduced by P. F. Jenkins, involves the *saddleback bird*, which is native to the islands of New Zealand. These birds tend to sing certain songs based on their territories. Birds in close territories sing similar songs, kind of like dialects for humans. These songs then get passed down from generation to generation. It is their culture. That said, there are instances where birds will mess up songs by adding a note, singing in a different pitch, or even replacing part of a song with another song. At this point, what one bird thinks is the "right" song may not be true for the neighboring bird![8] The transfer of ideas from one entity to the next is not perfect, yet life goes on.

A final example will help illuminate the complexity of transferring ideas. Look at the picture on the left. What do you see? Do you see a young woman turning her head away? Or do you see an old woman looking down? It is possible you see both. Now, look at the dress. What color is the dress?* I see black and blue. Others see white and gold. No matter how hard I try, I will never be able to see a white and gold dress.

Figure 29. What do you see on the left? What color is the dress on the right?

Now, imagine you sent one of these images to someone else. You see one thing and the other person sees another thing; however, you are both looking at the same thing! In a similar way, ideas can often be misconstrued based on the priming of an idea space. When two people may think they are taking about the same thing, it is entirely possible that they have two distinct perceptions

---

* If you are not familiar with the famous dress and you are currently viewing the image in black and white, then I recommend searching the internet to experience its true colors.

of that singular idea. In some instances, two people might eventually be able to see what the other person is talking about, like the young/old lady. In other instances, two people might never be able to understand one another, like the dress, which took the internet by storm. Clearly, there is more than meets the eye when transferring ideas.

As you read this book, you have probably developed some sort of preliminary understanding of what an idea space is. However, your perception of an idea space is completely different than someone else's perception! That said, it does not mean that there is not a base level of knowledge between all of us.

KOANS AND THE IDENTITY OF AN IDEA SPACE

While transferring idea spaces can often be complicated, there are instances where the transfer is almost perfect. In such cases, you have encountered a koan: a phrase, story, dialogue, question, statement, picture, feeling, or sensation that perfectly encapsulates the identity of an idea, or a moment in space and time. The identity is the pure essence of a particular idea (figure 30). It is a special idea that succinctly captures the whole of an idea space. This does not only include the thoughts of an idea space. The koan encompasses the emotions, sensations, and perceptions of the idea space as well. Koans are important, because they are vital in building your principles, as we shall see in Chapter 9.

Figure 30. The identity of an idea.

The word *koan* is Japanese for "public case." It is a tool Zen masters use to see whether a student has achieved a certain level of understanding, or awakening, in regard to an ineffable truth of our world. The koan experience is usually nonverbal. To demonstrate the ineffable nature of a koan, here is an example from *The Gateless Gate*, by Mumon Ekai:

WELCOME TO THE IDEA SPACE

*A monk asked Joshu, a Chinese Zen master: "Has a dog buddha nature or not?"
Joshu answered: "Mu" [No-thing.]*

Breaking koans down more practically: Have you ever thought about something, then, as you go through life, you come across someone or something that perfectly captures your thought? A lightbulb moment that leaves you speechless? A brain blast? In those instances, you have come across a koan. A moment of awakening. It is best to write down your koans, as they are unique to you and vital in shaping the large-scale structure of your idea space, as we shall see later in the book.

Let's look at a specific example. Let's say we have an idea space with various subideas within it: fear, impermanence, courage, and freedom. As time passes, the idea space naturally changes form. At a random point in time, a wild koan appears, capturing the pure essence of impermanence (figure 31)! For instance, notice what sensations arise when you read the following Heraclitus quote: "No man ever steps in the same river twice, for it is not the same river and he is not the same man." If that does not resonate with you, then this one may: "Whatever has the nature to arise has the nature to pass away."

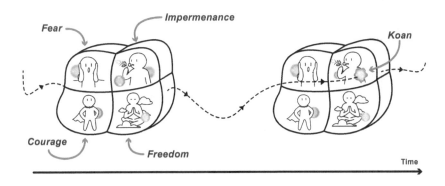

Figure 31. Koans capture the identity of an idea space.

These two quotes are koans that perfectly capture the concept of impermanence. They are the identities of that idea space. They are what you will remember when thinking of or discussing impermanence. Of course, an idea space can have many identities, such as this one on impermanence.

LIFTING THE FIRST VEIL OF ILLUSION

The first veil of illusion lifted in this book reveals the concept of an idea space (figure 32). Prior, your understanding of consciousness may have been as an ineffable, abstract object. Now, we are defining consciousness more properly by giving it some rigid structure, so you can view your thoughts, emotions, sensations, and perceptions as objectively as you would view objects in spacetime—devoid of "I."

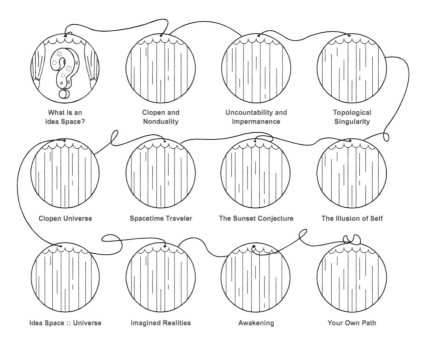

Figure 32. Lifting the first veil of illusion on the Path of Awakening.

To recap, your idea space consists of your thoughts, emotions, sensations, perceptions, and the empty set. Your idea space is unique to you, uncountable, has zero measure, and sits at the center of your own observable universe. The combination of idea spaces between people gives rise to macro idea spaces, and the transfer between idea spaces is seldom perfect. However, there are times when a seamless transfer takes place, at which point you have encountered a koan. In the coming chapters, we shall take a deeper dive into some of these properties to continue lifting the veils of illusion.

Chapter 2
NONDUALITY

"There's so much everything that nothing is hidden quite nicely."
- Wislawa Szymborska

Begin by taking a couple of deep breaths to yourself. It does not have to take long. Simply bring awareness to the here and now. Deep inhale in . . . and slow exhale out . . . Take the next 20 seconds to label ideas as they show up in your idea space. This is a thought. This is an emotion. This is a sensation. Simply look away from the book, take a moment for yourself, and note what arises.

The practice of mindfulness is perfectly summed up by the following story. One day, a man heard of the Buddha's great wisdom and set out to seek his teaching.* After a long journey, he found the Buddha while he was collecting alms. The man asked the Buddha for teaching, but the Buddha requested he wait until the end of his alms round. After traveling a great distance, the man was in no mood to wait. After his third request, the Buddha gave in and taught the man as succinctly as possible, "In the seeing, there is only the seen. In the hearing, there is only the heard. In the sensing, there is only the sensed. In the thinking, there is only the thought."[10] These phrases capture the pure essence of mindfulness.

The Buddha's teaching highlights the importance of groupings as the most foundational form of knowledge. By categorizing our experiences into distinct groups, we can better understand and navigate the complexities

---

*Siddhartha Gautama, known as the Buddha, was a historical figure who originated from India around the 5th to 4th century BCE.

of our world. For instance, think about organizing a library filled with books. Before you can find a specific book or understand its content, you may start by grouping the books by genre or subject. By categorizing them into sections such as fiction, non-fiction, self-help, or science fiction, you create a system that helps you navigate and locate books based on their shared characteristics. This process of grouping is foundational—it helps us understand the layout and content of the library even before we engage in further quantification, like counting how many books there are in each category

In turn, many practices begin by having students label their experience as it arises. This is a thought. Here comes the breath. That is a touch sensation. This a smell. That is a sound. There is a squirrel. In other words, you group and label your experience in terms of elementary ideas of thoughts, emotions, sensations, and perceptions as they come and go.

The catch is you can never cling onto a specific idea, because it is impermanent. It would be like trying to hold water in your hands. As soon as you try to grab it, it changes shape into something else. The key to mindfulness is simply to see the arising of an idea, the passing of an idea, and both the arising and passing of an idea. That is it.

The conundrum is everyone groups ideas differently. For instance, how would you group the items in the below image?

Figure 33. How would you group these?

If you choose to do it by shape (triangle, rectangle, and circle), notice someone else could have done it by food vs non-food, or vice versa. Similarly, elementary ideas of thoughts, emotions, sensations, and perceptions can be grouped in various ways by different people. For example, where you start and end your breath may be different than where someone else decides to start and end their breath.

To provide clarity and ensure that we are talking about the same ideas, we need to develop a ubiquitous structure that transcends personal idea spaces. In other words, we need to develop a language that ensures we group ideas

NONDUALITY

in a common way. We can achieve a shared foundation of comprehension through the groupings of closed, open, and clopen.

A *closed idea space* is similar to a closed mindset. All the ideas are present. For example, a closed idea space can be found in an "expert" who believes he or she knows everything there is to know on a particular subject. No new ideas will enter the idea space. An *open idea space* is analogous to an open mindset. For every idea that is had, there will always be a new one in the vicinity. An open idea space can be found whenever someone is learning something new for the first time. In these instances, a million new questions arise whenever a topic is discovered. A *clopen idea space* is an idea space that is fully open and fully closed at the same time. It provides a framework for *nonduality*, or when two seemingly opposite ideas are simultaneously true. A clopen idea space can occur in an individual who has an understanding of a topic, yet is open to new knowledge on the subject, even if the new knowledge contradicts their current understanding or tells the opposite story.

Building a clopen idea space is vital to living a fulfilling life. Clopen shows you two opposite ideologies can be simultaneously true. As Niels Bohr, the father of the atom, said, "The opposite of a correct statement is a false statement. But the opposite of a profound truth may well be another profound truth." Clopen is necessary now, because we live in a polarized era where stress is inescapable, and people are increasingly skeptical of one another based on very limited information, such as what political party someone aligns with or what neighborhood someone is from. Clopen breaks you free from the story you have been telling yourself and opens your heart by painting an objective picture of reality where true and false live harmoniously.

To unpack clopen, we shall first take a deeper dive into what it really means for an idea space to be closed and open. Then, we'll lift the next veil on our journey to show your idea space can be simultaneously closed and open.

CLOSED IDEA SPACES

On one hand, a grouping is considered closed if it contains all its limits.[*] For example, imagine we have a disc whose radius is one (figure 34-a).

---

[*] See Idea Space Whitepaper at www.TheIdeaSpace.io for more a formal approach to limits, open, closed, clopen, and the concept of an idea space.

This disc is considered closed, because it contains all the points inside the disc and the points on the boundary. Similarly, the line segment [0,5] is considered closed, because it contains all the points within the interval *and* the points zero and five on the boundary (figure 34-b).

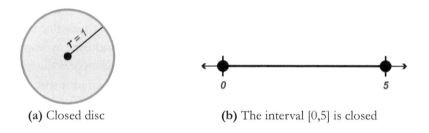

(a) Closed disc   (b) The interval [0,5] is closed

Figure 34. Two different representations of closed.

An idea space is closed when it contains all its limits (figure 35). This can be interpreted as, "Hey—I know everything there is to know about this." It can also be when you are simply focused on a specific task. In other words, every thought, sensation, emotion, and perception in a closed idea space is self-contained and can be reached. This is not necessarily a "bad" thing. Sometimes, it is important to have closure or the ability to focus on what is already present instead of continuously jumping to new ideas. For example, as you finish up a project, it is best to bring the story full circle and avoid introducing new material.

Figure 35. A closed idea space.

To experience this *focused awareness,* turn your attention to the sensations in your feet on the ground. Pick a point where you feel it most, and focus all your awareness there. Lift your toes and notice the pressure applied to the

NONDUALITY

bottoms of your feet. Close out all other experiences. Attune your attention to one point and one point only. What do you notice?

In a closed idea space, you only notice one sensation—your feet on the ground. That said, while your focus was there, other sensations were still happening, like the feeling of your clothes touching your body. But, since you focused your attention on one spot and one spot only, then those other sensations were shut out. You experienced a closed idea space.

OPEN IDEA SPACES

On the other hand, an open grouping does not contain all its limits. For example, imagine we have the same disc with a radius of one (figure 36-a). This time, the disc is considered open, because it contains all the points inside it, but the disc does *not* contain the points on the boundary. Similarly, the line segment (0,5) is considered open, because it contains all the points within the interval, but does *not* contain the points zero and five on the boundary (figure 36-b). In both these instances, you get closer and closer to the boundary, but never actually reach it. As you approach the boundary, there is always a new point that appears. Simply put, the boundary, or limit points, are not included in the grouping.

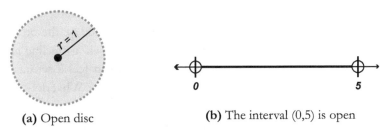

(a) Open disc   (b) The interval (0,5) is open

Figure 36. Two different representations of open.

A great illustrative example of the differences between open and closed can be found in video games. If you have ever played one, then you will be familiar with the concept of lag, or glitch. For instance, a character can sometimes be stuck facing a wall and endlessly walking toward that wall. He or she never fully gets to the wall, but they are still walking toward it. That is open. If they go—splat!—and finally reach the wall, then that is closed.

An idea space is open if it does not contain all its limits (figure 37). This is similar to an open mindset. You can have many thoughts, emotions, sensations, and perceptions present within your idea space, but there are some ideas that are unreachable, like the wall in a video game. In a sense, as you approach a specific elementary idea, there is always a new idea that populates the idea space. In the same way that a closed idea space is not necessarily "bad," an open idea space is not necessarily "good." For example, when working with people, it is important to set clear, bounded goals to limit indecisiveness and achieve results.

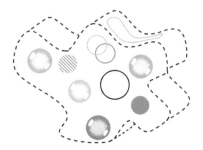

Figure 37. An open idea space.

To experience this *choiceless awareness*, simply focus on whichever sensations arises. This could be your feet, your hands touching this book, or your tongue on the roof of your mouth. Let the joyride of the breath take you to whatever sensation arises. Do not limit yourself to feeling any one bodily sensation in particular. Instead, see if you can feel a sensation you have never noticed before, like your eyes in their sockets. What does that feel like?

In an open idea space, you are not tied down to a particular sensation and you notice sensations as they arise. The trick is as soon as you notice a particular sensation, a new sensation appears elsewhere.

CLOPEN IDEA SPACES

Unlike doors, idea spaces can be both simultaneously open and closed, or clopen. Here, we lift our second veil of illusion. Zen master Seung Sahn Sunim once said, "There is no right and no wrong, but right is right and wrong is wrong." With clopen, there is no open and no closed, but open is open and closed is closed. All in all, clopen provides the necessary framework for two seemingly opposite ideas to be simultaneously true.

From the standpoint of consciousness, clopen directly relates to nonduality. As Alan Watts said, "Duality arises only when we classify, when we sort our experiences into mental boxes, since a box is no box without an inside and an outside."[11] In other words, when we group items we get a duality: open vs. closed; good vs. bad; fast vs. slow; etc. However, if we take a step back and observe the situation as a whole, we see there is no duality—the situation is clopen: open and closed; good and bad; fast and slow; etc.

In a way, clopen allows you to not get attached to a particular idea by showing you the opposite viewpoint might be just as valid. What is slow without fast? What is good without bad? Whenever we group, we tend to forget we are creating a duality, which sets us on a path of attachment to a particular way of thinking. Then, we develop a certain perspective of the world that is unique to us and may be completely different than someone else's perspective. In each case, both viewpoints could be equally true or equally false. By embracing the concept of clopen and recognizing the validity of seemingly opposing perspectives, we can free ourselves from the grip of attachment to one particular point of view and find greater peace amid the ever-changing nature of ideas and experiences.

A prime example of a clopen phenomenon is the relationship between mind and body, or *nāmarūpa* (Pali). Some philosophers have suggested this relationship will only be understood with reference to concepts that are in some way neutral.[12] Therefore, the answer to "Is life controlled by the mind or body?" is both—the situation is clopen.

To better understand clopen, you must first understand *complements*. Let's look at the concept intuitively. A complement is what remains in a set when one takes part of the original set out. For instance, imagine we have a set of four pears and three apples (figure 38-a). Then, I take out three apples, and I am left with four pears. The complement of the three apples is the four pears; and, the complement of the four pears is the three apples (figure 38-b). The set of apples and the set of pears are complements of one another, as they make up the whole set when they are together.

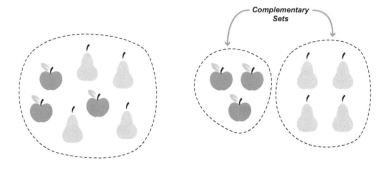

(a) A set of apples and pears

(b) The sets of apples and pears are complements of each other

Figure 38. Apples and pears are complements as they make up the complete set together.

From here, we can explore the relationship between open and closed more closely. Namely, *openness is dual to closedness*: the complement of an open set is closed and the complement of a closed set is open. To illustrate this point, let's explore two examples leveraging our discs and lines from before. As with most things in life, *understanding the pictures is easier than understanding than words*.

Let's start with our disks and imagine they are a part of a larger, arbitrary space that is filled called M. Within M, let's assume we have an open disc, called O (figure 39-a). "Open" means our disc contains all the points within it, but does not contain the points on the boundary. Now, if I were to take out the open disc, O, from the space M, then the resulting border within M would be closed (figure 39-b). Why? Because the complement of an open border is a closed border. Since we took out a disc whose border is open, then what remains must be closed.

Similarly, let's assume we have a closed disc, called C, within the same space M (figure 39-c). "Closed" means C contains all the points within it *and* all the points on the boundary. In this case, if I were to take out the closed disc, C, from the space M, then the resulting border within M would be open (figure 39-d). Why? Because the complement of a closed border is an open border. Since we took out a disc whose border was closed, then what remains must be open. Openness is dual to closedness.

NONDUALITY

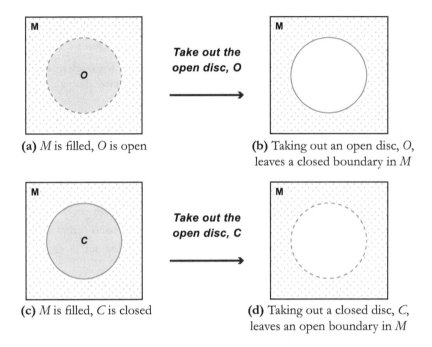

Figure 39. (a-b) Taking out an open set, *O*, of space *M* leaves a closed boundary in *M*. **(c-d)** Taking out a closed set, *C*, of space *M* leaves an open boundary in *M*.

For cognitive ease, let's do a second example. Bring back to mind the closed line interval between the points zero and five. If I take out the *open* interval (1,2), then what remains is the union [0,1] and [2,5]. Since one and two were not included in the open interval I took out, then one and two must exist in the remaining interval. The resulting border is therefore closed (figure 40-a-b).

Symmetrically, if I have the same closed interval [0,5] and I take out the *closed* interval [1,2], then what remains is the union of [0,1) and (2,5]. Since one and two *were* included in the closed interval I took out, then one and two must *not* exist in the remaining interval. The resulting border is now open (figure 40-c-d). Openness is dual to closedness.

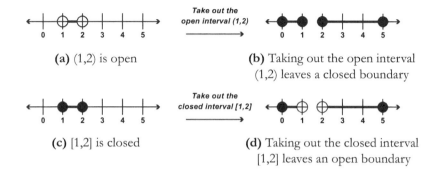

Figure 40. (a-b) Taking out the open interval (1,2) produces a closed boundary in the remaining set. **(c-d)** Taking out the closed interval [1,2] produces an open boundary in the remaining set.

Here we are able to see more clearly that the complement of a closed set is open and the complement of an open set is closed. With this understanding established, we can finally get to clopen by answering the profound question: *What is the complement to everything?*

Imagine we have a space, M, that contains everything. It is considered closed because it contains all its limits. In other words, since it contains everything, there is nothing that would not be in M. Hence, it is closed.

Now, the complement of everything is nothing, or the empty set, $\emptyset$. The empty set is also closed since it has no limits to even worry about. Thus, M and $\emptyset$ are both closed. Their complements, $\emptyset$ and M, respectively, are therefore open. M and $\emptyset$ are both closed and open. In other words, everything and nothing are both clopen.

To illustrate the point, imagine taking nothing out of space M, which contains everything. Since M is closed and $\emptyset$ is closed, M becomes open (figure 41-a). But, since you took nothing from M, then it remains closed (figure 41-b). Thus, M is clopen. Furthermore, since nothing is the complement of M, then nothing must also be clopen. Simply put, M and $\emptyset$ are both closed and open.

NONDUALITY

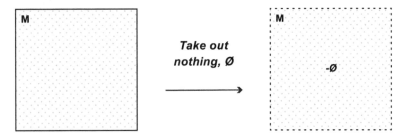

(a) Taking out nothing, ∅, from a closed set produces an open set

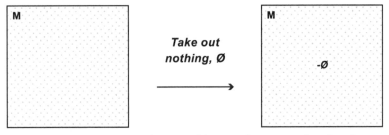

(b) However, you took out nothing, so the set remains closed

Figure 41. The empty set, ∅, and the space, M, are both clopen.

In terms of idea spaces, you can get from a closed idea space to an open idea space by taking out nothing, ∅ (figure 42). In other words, you can shift your attention between a closed and open idea space at will. This is important, because it allows you to cultivate a sense of flexibility and adaptability in your thinking. You can then approach problems and situations from different perspectives by considering the alternate viewpoint. Thus, helping you reduces stress and anxiety by allowing you to let go of rigid, inflexible thinking patterns and be more open to new possibilities. The beauty of clopen is you do not have to change anything, since the relationship between a closed idea space and an open idea space is nothing!

Think back to the sensation exercises we did in the closed and open sections where you focused on the foot with focused awareness, then the body with choiceless awareness. Although the baseline experience in sensations was the same, you experienced a closed idea space in the first instance and an open idea space in the second. What changed in between both? Nothing! Through mindfulness, you are able to shift your awareness and effortlessly change the configuration of your idea space. All without changing a thing.

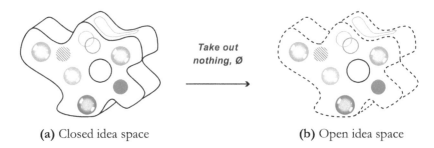

Figure 42. Take out nothing, ∅, from a closed idea space to achieve an open idea space.

THE BEAUTY OF CLOPEN

When we group objects, which we do every second of our lives, we do so arbitrarily. For instance, what you deem fun may be different than what I deem fun. After a grouping is made, you may tend to think your grouping is "right," but it may be completely different than someone else's grouping. Clopen reduces the urge to cling onto a right or wrong view of the world by understanding that many things in life can simultaneously be both right and wrong. It teaches us to view things for what they are instead of the meaning behind them.

Prime examples that illustrates the clopen nature of our minds are two koans found in the Zen classic, *The Gateless Gate*. The title of the book is itself clopen: gateless gate. In these koans, the nature of buddha, or an awakened state of mind, is questioned:

<u>Koan 1</u>
Daibai asked Baso: "What is buddha?"
Baso said: "This mind is buddha."[13]

<u>Koan 2</u>
Another monk asked Baso, "What is buddha?"
Baso replied, "Not mind, not buddha."[14]

So, which is it? Is the mind buddha? Or, is the mind not buddha? Both are true. The mind is buddha. The mind is not buddha. The situation is clopen. Understanding the clopen, or nondual, reality of nature is one of the key tenets of Zen. Thus, we are able to lift the second veil of illusion on our Path of Awakening. Prior there was only open versus closed. Now, there is clopen: open and closed (figure 43).

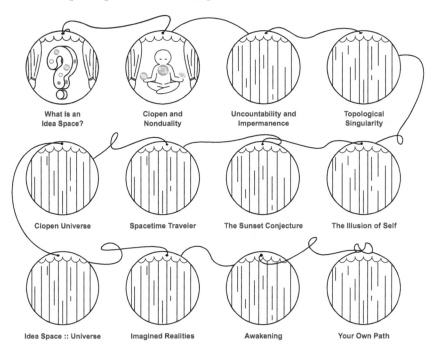

Figure 43. Clopen and nonduality represents the second veil of illusion on your Path of Awakening.

In the previous chapter, we introduced the idea space as a scientific model for your mind. Now, we have explored the different ways of grouping thoughts, emotions, sensations, and perceptions into closed, open, and clopen idea spaces. Neither one of these groupings is "good" nor "bad." They all have their place in time. The beauty of clopen, though, is it allows for two opposite ideas to be simultaneously true: open and closed, right and wrong, black and white, true and false.

Clopen is important, because it helps prevent clinging, which is the main source of suffering. Cling to unpleasant objects and you will be unhappy. Cling to pleasant objects and you will be unhappy once they are gone. Through clopen, we see that the idea we choose to attach ourselves to has a counterpart that may be equally as valid.

In life, clopen provides a neutral approach to the conundrum of mind-body duality: Is life controlled by the mind? Or is life controlled by the body? With clopen, the two coexist in peace: Life is controlled by both the mind and the body.

Another, more physical, example of clopen lies in the concept of *particle-wave* duality. In quantum mechanics, light can act either as a particle or as a wave. When simply observed, light acts like a wave. Make a precise enough measurement and light acts like a particle instead. So, which is it: is light a particle or a wave? Both. The issue is clopen. More on this in Chapter 8.

In the day to day, clopen can be found whenever you have conversations with people. For instance, one person may say something you disagree with, and you get into an argument. In those instances, know you may both be right and both be wrong. Like the young/old lady and the blue/gold dress in the first chapter, it all depends on your perspective. There is no right and no wrong, but right is right and wrong is wrong.

A great way to remind yourself about the clopen nature of reality is to perform a mindful pause, like S.T.O.P.P. In this instance, you attune your attention to nothing, Ø, which is clopen. Thus, as you clear your mind, you are forced to see everything—every thought, emotion, sensation, and perception—since everything is the complement to nothing. Or, as poet Wislawa Szymborska says, "There's so much everything that nothing is hidden quite nicely." During this brief pause, you have the opportunity to start fresh with beginner's mind and eliminate any biases as you perceive the nondual reality of the world.

Overall, groupings are the most foundational form of knowledge that builds our idea space. It supersedes the need to count or even make measurements.[15] Thus, grouping, whether of thoughts, emotions, sensations, perceptions, or objects, is the most elementary form of human experience. Then, mindfulness is the simple act of noting when groupings arise, when groupings pass, and the overall arising and passing of groupings.

My challenge for you as you go about your day is to notice when you group anything you hear, see, feel, think, or perceive. For instance, "That is wrong." "There is a book." "I hate this." You may be surprised as to how often you group objects. In those instances, ask yourself:

Is there a hearer doing the hearing [or, seeing, feeling, perceiving]?
Or, is there simply the act of hearing [or seeing, feeling, perceiving]?

Chapter 3

IMPERMANENCE

"Nature does not hurry,
yet everything is accomplished."
- Lao Tzu

Take a second for yourself. Breathe deeply in and out your nose . . . Take the next 20 seconds or so to see if you can count all your thoughts, emotions, sensations, and perceptions. Do not let yourself be fooled by your ideas. Some are very soft whispers, like "Hey—it's quiet in here." Others come from behind and say, "There haven't been many thoughts yet, have there?" Take a moment to pause. Can you count all your ideas?

Impermanence dictates the fact that the world is always changing. Even now, the Earth is spinning around its axis at 1,000 miles per hour (mph), rotating around the sun at 66,600 mph, and rotating around the center of our galaxy at around 514,500 mph (figure 44). The universe is in a constant state of flux.

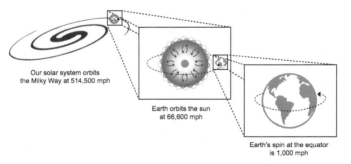

Figure 44. The universe is in a constant state of flux.

Since the universe is in a constant state of change, it is no surprise your idea space is also impermanent (figure 45). Although you can have similar thoughts from one moment to the next, no two thoughts are the same, just as no two breaths are the same. Simply put, you are not the person you were five years ago and you are not the person you will be five years from now.

Figure 45. Your idea space is impermanent, or uncountable. It is always changing.

Understanding the impermanent nature of our idea space prevents us from clinging onto any one idea. This is vital, because clinging is the main cause of suffering. For example, if someone clings to the idea they need to have a certain job or possess a certain material item to be happy, then they will constantly be in a state of craving and dissatisfaction. If they are unable to obtain that job or item, then they will suffer. Even if they do obtain it, the satisfaction will be temporary and they will soon find something else to cling to, perpetuating the cycle of suffering.

On a more precise level, impermanence is synonymous with *uncountability*. For instance, according to general relativity, space and time are both uncountable. For every measurement you make, even an infinite amount, I can always make a measurement you did not make. Similarly, an idea space is uncountable: For every idea you have, I can always list an idea you did not have. As we shall uncover, uncountability defines a size to infinity, which is larger than the infinity we use to count. This is the key to lifting the next veil of illusion on our Path of Awakening. Prior there was only infinity. Now, there are different degrees of infinity.

Of course, don't take my word for it. Test the fact your idea space is uncountable for yourself. Were you able to count every idea present in the opening of this chapter? Even now, are you able to count every thought, emotion, sensation, and perception present? Did you notice the sensations in your left foot before this sentence? How about your eyes as you blink? The sounds around you? These sensations are always present, but we are seldom aware of them all at once. As soon as you become aware or count

IMPERMANENCE

an idea, a new one appears. Thus, your idea space is impermanent, or uncountable.

Although the brain naturally filters out much of the sensory input it deems as noise, the impermanent and uncountable nature of the idea space still remains. Even if we are not consciously aware of every sensation or thought, they are still present and constantly changing, making our idea space impermanent and uncountable.

This chapter is dedicated to demonstrating how your idea space is uncountable, or impermanent. We'll start by reviewing what it means to count with our numbers and basic functions. This will lay the necessary foundation to understand what it actually means to "count". Then, we'll look at why spacetime is uncountable. This serves as a good reference point as to what it means for something to be uncountable. Finally, we'll see why your idea space is uncountable. The crux of the argument lies around the question: *Did gravity exist before Isaac Newton brought forth the idea of gravity?*

Before we start, ask yourself: How often do you count in a day? Fifteen minutes until the end of this call. Twelve more miles until I turn. Eleven more reps. Thirteen pages left to read. Once we create a grouping, the next fundamental stage of being human is to count how many objects are in the grouping. Take note as to how often you count in a day. It may be more than you think.

NUMBERS

Understanding numbers is crucial in understanding what it means to count, an activity we mindlessly do 24/7. Overall, there are four main types of numbers:

Natural Numbers, $\mathbb{N}$ = 1, 2, 3, 4, 5, 6, … | Whole numbers whose value is greater than 1.
Integers, $\mathbb{Z}$ = …, -2, -1, 0, 1, 2, 3, … | Positive whole numbers, zero, and negative whole numbers.
Rational Numbers, $\mathbb{Q}$ = 2/3, 5/4, 8/3, 9/10, etc. | The fraction between any two integers, such that the bottom integer is not 0.
Real Numbers, $\mathbb{R}$ = 2.3432…, or 231.12312…, or 2.97631…, etc. | Any decimal expansion that never terminates (it helps to picture a number line).

The main distinction between the rational numbers, ℚ, and the real numbers, ℝ, is certain numbers, like pi, π, are nowhere found in the rational numbers, because one cannot divide any two integers to create pi. So, we need to introduce the real numbers to capture the complete number line.

There is a natural nesting between these four sets. The natural numbers, ℕ, are a subset of the integers, ℤ, which are a subset of the rational numbers, ℚ, which are a subset of the real numbers, ℝ (figure 46). We've omitted the complex numbers, which all these other numbers are nested under, since they are not necessary for the purposes of this discussion.*

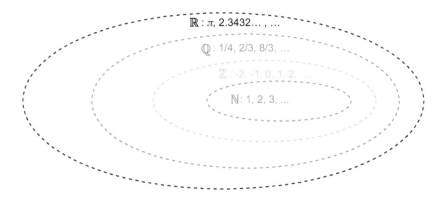

Figure 46. The nesting between different types of numbers.

FUNCTIONS

Understanding functions allows us to better see how numbers relate to the objects around us, and it is the final ingredient we need to comprehend what it means to "count." Functions, f, are imaginary machines that transform any element x in set A to an element $f(x)$ in another set B (figure 47-a). Functions also go by many other names such as maps or transformations. For example, I can have a function that transforms a dog into a cat (figure 47-b).

---

* Fun fact: the complex numbers are algebraically closed, while the other numbers are algebraically open. This means the set of complex numbers houses a solution to any polynomial, while the other number systems do not.

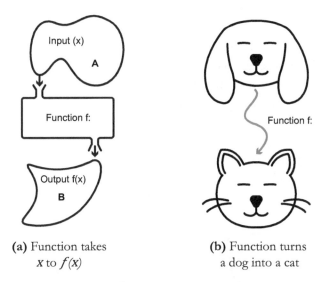

(a) Function takes x to f(x)

(b) Function turns a dog into a cat

Figure 47. A function, f, transforms an input x that lives in A into an output $f(x)$ in B.

There are many types of functions in the world, but I want to focus on three here: injective, surjective, and bijective. An *injective*, or one-to-one, function means every element in A is mapped exactly once to an element in B (figure 48-a). Note: not all of A has to map onto all of B. A *surjective*, or onto, function means all of A maps onto all of B (figure 48-b). Here, we see multiple elements, say a and a', can map to the same element, b. Lastly, a *bijective* function is both injective and surjective, or both one-to-one and onto. It maps all of A onto all of B exactly once (figure 48-c).

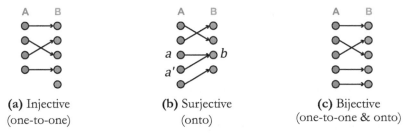

(a) Injective (one-to-one)

(b) Surjective (onto)

(c) Bijective (one-to-one & onto)

Figure 48. The three main types of functions: **(a)** injective, **(b)** surjective, and **(c)** bijective.

CARDINALITY AND COUNTING

Two sets are said to have equal *cardinality*, or the same number of elements, if and only if there is a bijection between both sets. The "if and only" signifies the reverse is also true: If there is a bijection between two sets, then the two sets have equal cardinality.

As a cute example, let's say I have three dogs here and three cats there (figure 49). I can map each dog to each cat exactly once (i.e., one-to-one). Plus, all the dogs map onto all the cats (i.e., onto). Since this function maps all the dogs onto all the cats exactly once, there is a bijection between the two sets. In turn, the two sets have the same number of elements, or equal cardinality.

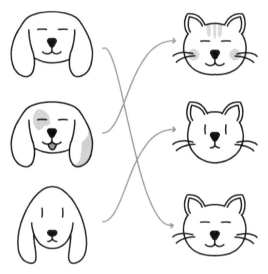

Figure 49. A function maps all the dogs to all the cats exactly once. Therefore, since there is a bijection, both sets have the same number of elements, or equal cardinality.

Let's look at some different types of cardinalities. An arbitrary set, S, is:

Finite if it is empty, $\emptyset$, or for some n in the natural numbers, $\mathbb{N}$ (i.e., 1, 2, 3, 4, ..., n).
Infinite if not finite. In others words, something is infinite if it goes on forever without bound, limit, or endpoint.

IMPERMANENCE

For example, the natural numbers, ℕ, are infinitely long. You can keep counting forever without bounds. Yet, I am still able to pick a finite natural number, like the number of atoms in the universe, 10^{80}. Even though the number is massive, it is still finite.

Furthermore, a set *S* can be:

Countable if it is finite or has the same number of elements as the natural numbers, ℕ.
Uncountable if it is not countable. As we shall see, this represents a set of infinity larger than the infinity we use to count.

Infinity and countability are not mutually exclusive. In other words, since there are infinite natural numbers, ℕ, you can count to infinity. An example will help illuminate. How many oranges are there in the set S (figure 50).

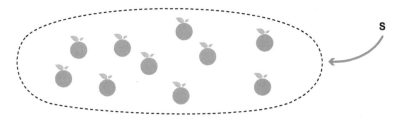

Figure 50. How many oranges are in this arbitrary set, *S*?*

In your head, you probably did something like figure 51. You mapped the natural numbers one by one onto our oranges. One orange, two oranges, three oranges . . . You'll agree, I could add as many oranges as I want to—an infinite amount even—and you could continue the pattern to count all the way until infinity. We can now properly define what it means to "count":

Counting means there exists a bijection between the natural numbers, ℕ, and whatever set we want to count.

In this case, we're counting good ol' oranges.

*Ten.

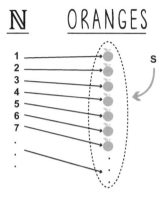

Figure 51. To count something means that there is a bijection, or one to one and onto function, between the natural numbers, $\mathbb{N}$, and a set.

Something is uncountable if it is not countable. So, uncountability means there is *no bijection* between the natural numbers, $\mathbb{N}$, and whatever set we want to count. Uncountability provides a size to infinity that is larger than the infinity of the natural numbers, $\mathbb{N}$.*

SPACETIME IS UNCOUNTABLE

Now that we have a better understanding of what it means to count, let's see what it means for something to be uncountable through a tangible example in spacetime.

According to general relativity, space and time are based on the real numbers, $\mathbb{R}$. Therefore, space and time are uncountable: For every measurement you make in spacetime, I can always make a measurement you did not make. This argument is dubbed *Cantor's Diagonal Argument* for reasons we shall illustrate through a fun, short story.

---

* Uncountability was developed by Georg Cantor (1845-1918) when he created one of the most beautiful theorems in human history: *the real numbers, $\mathbb{R}$, are uncountable*. Also, fun fact: $\mathbb{N}$ has the same cardinality, or number of elements, as $\mathbb{Q}$ and $\mathbb{Z}$. This is weird, because you would think the ratio of $\mathbb{Q}$ or $\mathbb{Z}$ to $\mathbb{N}$ is 2:1. See whitepaper for this proof and a more rigorous proof as to why $\mathbb{R}$ is uncountable.

Joe and Misty are two best friends who love science. One day, they decide to test whether space and time are uncountable. To do so, Joe buys a special ruler called *Deep Rule* he found in the magnificent encyclopedia of *The Hitchhiker's Guide to the Galaxy*. This toy ruler is able to perform computations of infinite complexities and make any measurement possible.

The instructions read as follows: "Dear life form, thank you for choosing Deep Rule. Please place the ruler on what you would like to measure and Deep Rule will count every measurement."

Joe takes the toy out of the bag and places it in front of him. He turns the machine on and instructs it to count all the measurements in a one meter length. A powerful laser shoots out, and within seconds, Deep Rule announces: "I have counted all possible measurements between zero and one meter. Please find the infinite list of measurements below:" (figure 52).

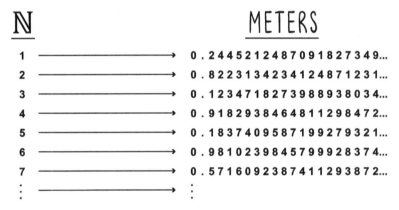

Figure 52. Deep Rule's infinite list of measurements between zero and one meters.

Misty looks at the screen and says, "Wait! Deep Rule, I think you missed a number." She then points at the screen and begins to describe a number that's not on the list.

"Deep Rule," Misty says, "what if I picked a number like: 0.3325178... ? In other words, the first decimal point in my number is different than the first decimal point in your first number. The second decimal point in my number is different than the second decimal point in your second number. The third decimal point in my number is different than the third decimal point in your third number . . ."

Instead of continuing on forever, she takes a red pen and draws it out on the screen (figure 53). She continues, "In other words, my number is different than every number in your list, because my number differs from each of your numbers in at least one position."

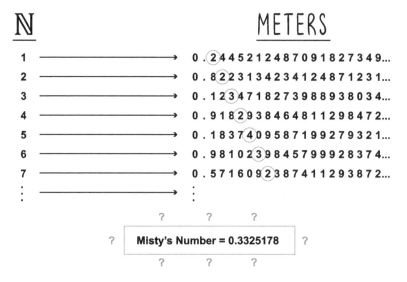

Figure 53. Where is Misty's Number in the infinite list of measurements Deep Rule counted?

"Deep Rule, you said you counted every measurement possible, but where is my number in your infinite list of measurements?" Misty asks.

Upon hearing this, Deep Rule responds, "Oh . . . no . . . Cannot . . . compute . . . Must . . ." and then spontaneously combusts into flames. The End.

This short story perfectly demonstrates why space is uncountable. If space were countable, then you could create a bijection between the natural numbers, ℕ, and all the spatial measurements, like Deep Rule did. However, using Misty's logic, which is Cantor's Diagonal Argument, she showed there was a measurement Deep Rule did not count in its infinite list! In other words, she chose a number that differs from every other number in the infinite list in at least one position. Therefore, the bijection between the natural numbers, ℕ, and measurements of space no longer holds, thereby making space uncountable.

You can repeat the same thought experiment with time to show time is uncountable. For example, if another machine called *Deep Time* counted every possible measurement between zero and one second, then I could always pick a time Deep Time did not count using the same logic. Namely, my measurement's first decimal point would not be the same as Deep Time's first decimal point in its first measurement. The second decimal point of my number would not be the same as Deep Time's second decimal point in its second measurement . . .

To better understand this, we turn to *reference frames*. Look straight ahead. That is your x direction. Now look to your right. That's your y direction. Finally, look up. That is your z direction. Together, these three dimensions of space (x, y, z) create your reference frame at a particular point in time (t). Then, whatever causes time to flow moves the reference frame through time (figure 54).

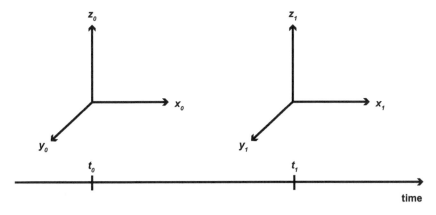

Figure 54. A reference frame is a point identified by three spatial coordinates (x, y, z) moving through time (t).

Now, if you want to measure something in the x dimension, then what would you do? I'd get out a ruler and get to measuring. Since this is a thought experiment, let's imagine we're using Deep Rule. Is 2.543 meters there? Yep, we got it. 1,203,021 meters? Sure, no problem. 1.6×10^{-35} meters? Why not. Essentially, everything we can measure in the x dimension can be measured using the real numbers, $\mathbb{R}$. Think of a giant, straight number line extending outward in front of you. Wherever you land, you'll find a point to measure.

Can you count every point in your *x* direction? No. There are an uncountable number of points there: For every point you pick, even an infinite amount, I can always pick a new point you did not pick. Look right. This holds true there too. Look up—still checks out. Now, take a deep breath. Focus on your breath and feel the passage of time. Yet again, that is uncountable. I can always choose a time you did not pick. That said, it does not mean that we can't pinpoint approximately how long ago the beginning of our universe was (around 13.8 billion years ago), but it is impossible to count all the different times that have existed since.

So, according to general relativity, space and time are uncountable, because you cannot map the natural numbers, $\mathbb{N}$ (1, 2, 3, . . .) bijectively onto the real numbers, $\mathbb{R}$ (2.342. . .). Simply put, for every measurement you make in spacetime, I can always make a measurement you did not make (figure 55). This is Cantor's Diagonal Argument.

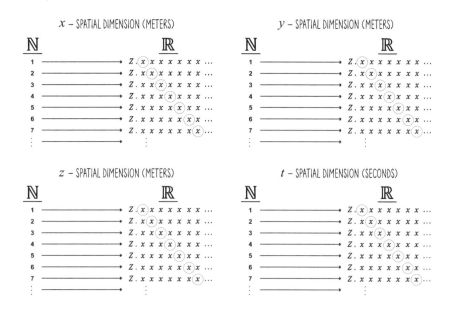

Figure 55. Cantor's Diagonal Argument for spacetime. *Z* represents any integer (e.g., … , -1, 0, 1, …) and *x* is any number between zero and nine.

Uncountability lifts another veil of illusion. Prior there was only infinity. Now, there are unique depths to infinity. For instance, the interval

[0, 1] has more points in it than the infinite amount of points in the natural numbers (1, 2, 3, 4, 5, …). So does (0, 0.5). So does (0, 0.00000001). And, as we'll see in the next chapter, even something with zero measure can have an uncountable amount of elements (figure 56). Uncountability can be found in the largest and smallest of sizes.

Figure 56. How can something be uncountably deep, yet have zero measure?

The uncountability of space and time directly translates to its impermanence. Explicitly, uncountability implies constant change. Even if you somehow were able to count all the points in the universe, there would always be an uncountable amount of points you did not count. As a result of this uncountable nature, we can infer a fundamental characteristic of the universe: it is in a constant, ever-evolving state of change. The most evident form of this lies in the mystery behind *dark energy*. Dark energy is the enigma responsible for the constant expansion of space. As we shall discuss later, this expansion is uniform at every point in space and time and is the underlying reason behind the impermanence of our universe.

THE IDEA SPACE IS UNCOUNTABLE

An idea space is uncountable, or impermanent, through a similar logic as to why spacetime is uncountable. The simplest way to test this is to sit for 20 seconds and count all the ideas that come and go. You'll notice for every idea you count, a new one appears! There will always be an idea you did not count.

Thus, the following proposition is stated: *Every solution to a problem raises new unsolved problems; whenever an idea is discovered, a new idea appears; a new discovery creates more questions than answers.*[16] Don't take my word for it. Test this idea out for yourself. When was the last time you discovered something new? Did it end right there? Or did the discovery lead to a million more questions about what you just discovered? If this proposition is true, then it provides the necessary framework for demonstrating why the idea space, or your mind, is uncountable.

For an illustrative proof, picture yourself living in the year 1680. You are tasked with counting all the ideas in the world. You go out and you count an infinite amount of them. Then, in 1687, Isaac Newton publishes *Philosophiæ Naturalis Principia Mathematica* and the "idea of gravity" comes into the picture. Where was the "idea of gravity" in all the idea spaces before? Was it in the first one? No. The second? No. It is nowhere found in the infinite list of ideas (figure 57). In other words, gravity did not "exist" until Isaac Newton brought forth the "idea of gravity". Similarly, the "idea of an idea space" did not "exist" until this book. Where was it in the infinite list of idea spaces beforehand?

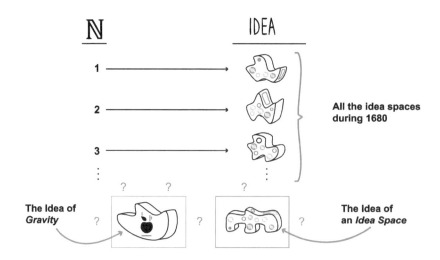

Figure 57. Where was the idea of "gravity" or the idea of an "idea space" in 1680?

This proposition lays the foundation for the idea space to be uncountable, or impermanent. There will always be new ideas to be had, even if there are already an infinite amount. Or, as John Wheeler, the father of black holes, said, "We live on an island of knowledge surrounded by a sea of ignorance. As our island of knowledge grows, so does the shore of our ignorance."

Furthermore, elementary ideas and fundamental ideas are also uncountable. This follows a similar logic as to why (0, 0.00000001) is uncountable. For instance, picture a dog in your head. Clearly, the dog is bounded. You can describe the dog to someone else as much as you'd like, but there will always be part of the dog that is not described. This can be a physical feature of the dog or an emotion the dog makes you feel. Even though ideas are bounded (you are imagining a dog, and only a dog), they can still be uncountably deep (you can never fully describe your idea of a dog).

Practically, it is impossible to count a specific thought, emotion, sensation, or perception. They're more like ineffable clouds whose impermanence we are forced to describe using words, pictures, songs, equations, tone, body language, or whatever other *transmitters* we can think of. This is the human experience. Hopefully, in the future, someone figures out how to transfer idea spaces like a Pokémon TM.*

What does this mean for me? Look forward in your x direction again. Take a moment to pause. What are you thinking about? Everything you're currently thinking, and have ever thought, is uncountable, or impermanent. It is not possible to count all the thoughts you had, the thoughts you are currently having, and the thoughts you will have. The best we can do is stay in the present moment and watch our thoughts as they arise and pass away. As Joseph Goldstein wisely states, "You experience the past as a thought in the present. You experience the future as a thought in the present."

Understanding the impermanent nature of our minds is liberating, and it lifts the third veil of illusion on our Path of Awakening (figure 58). Often times, we get caught up in a moment and are filled with negative thoughts or emotions. But those thoughts were not always there. At one point, they did not exist. Then they existed. And, as the Buddha said, "Whatever has the nature to arise has the nature to pass away." It is our task to notice the arising of an idea, the passing away of an idea, and both the arising and passing away of an idea.

Up to this point, we've been able to detach ourselves from our thoughts, emotions, sensations, and perceptions through the concept of

---

* TMs are items that can be used to teach Pokémon new moves or abilities, like cut, fly, or swim.

an idea space. We also saw how to view the world through an objective, nondual perspective through clopen. Now, we're able to see our mind as an ever-changing, impermanent landscape through uncountability.

After seeing impermanence, you understand you cannot cling to thoughts, emotions, sensations, or perceptions, because they are uncountable and bound to change. This counts for both pleasant and unpleasant ideas. That said, it doesn't mean we cannot enjoy the pleasant ideas when they arise. But, if we cling onto those pleasant sensations, then we'll be upset when they are gone. Developing a deep understanding of impermanence relieves this stress. In an impermanent world, we no longer need to attach to our thoughts, our emotions, or even our names and identities. We see the ever-changing world for what it is. As ancient Chinese philosopher Lao Tzu said, "Nature does not hurry, yet everything is accomplished."

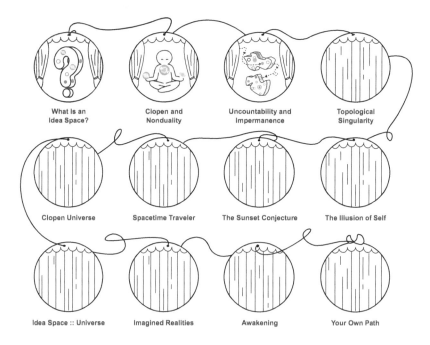

Figure 58. Uncountability and Impermanence lifts the third veil on our Path of Awakening.

Clinging is the main source of suffering. Through impermanence, we see clinging is simply impossible. As soon as you cling to one thing, it changes. Therefore, you are not bound to your preconceived notions of yourself nor of the world. In a way, you are always given the opportunity to reinvent yourself. No matter who you are. No matter where you are.

Chapter 4

MINDFULNESS

"There are only two ways to live your life.
One is as though nothing is a miracle.
The other is as though everything is a miracle."
- Albert Einstein

As we begin, let's do a mindful pause. Put your hand in front of your face and try to fog it up, as if it were a mirror or glass. Do this a few times. Now, perform the same action, except with your mouth closed—breathe in and out through your nose. Can you feel the deepness of this ocean breath? Can you send this breath deep into your lower belly? Simply sit with this breath as you continue to read this chapter.

Mindfulness is awareness that arises through paying attention, on purpose, in the present moment, nonjudgmentally; meditation is the formal practice of mindfulness; a practice is the embodiment of an approach to a concept.[17] In turn, your idea space is your object of meditation. The beauty of mindfulness is you can be mindful of anything—any thoughts, emotions, sensations, or perceptions present. Different experiences conjure up different mindful states, some more powerful than others.

Many people believe mindfulness to be emptiness of thoughts: *I am mindful when no thoughts are present.* Although emptiness is one aspect of mindfulness, it does not tell the whole story. As philosopher Sam Harris writes, "The problem is not thoughts themselves but the state of thinking without knowing we are thinking."[18]

The best way to develop a mindful lifestyle is to practice it formally through meditation. Then, you take the mental state you experience during meditation into your day-to-day life. Thus, you can become mindful while walking, eating, brushing teeth, using the bathroom, standing, sitting, writing, lying, breathing, and simply being. If we become mindful, even if only for a moment, throughout the activities that build our lives, then we're on the right path. As an Ancient once said, "Better than one hundred years lived without seeing the arising and passing of things is one day lived seeing their arising and passing."[19]

Although there are many different types of mindfulness practices, for beginners I recommend author Diana Winston's model for working with the *Spectrum of Awareness*. This model gives us a map to understand the different ways to meditate with awareness. Again, it is only a map—your mind and body are the territories.

The Spectrum of Awareness shows how the various forms of mindfulness fit within one another, so you can have flexibility in your meditation practice. Winston breaks the spectrum into four categories: (a) focused awareness, (b) investigative awareness, (c) choiceless awareness, and (d) Selfless awareness (figure 59). The spectrum is further split into object-based awareness and natural awareness. The first three categories are object based (e.g., breath, sensations, thoughts, emotions, sounds, etc.), while natural awareness is objectless (i.e., awareness aware of itself). No one part of the spectrum is better than the other. You'll use each type depending on the circumstance you're in.

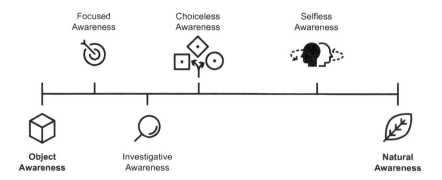

Figure 59. The spectrum of awareness.

Focused awareness is similar to a closed idea space. It is awareness tuned into a specific part of the idea space (figure 60-a). In a sense, it is like taking a picture of a specific object. It requires effort and is great when you are distracted. To give this a try, attentively focus on your breath for 20 seconds. If you get lost in thought, simply come back to the breath.

Investigative awareness is guiding your awareness to various objects within your closed idea space (e.g., breath, thoughts, etc.) (figure 60-b). In a sense, it is like taking a landscape photo of something. One can experience this through a body scan meditation. Start with your toes and slowly move awareness up your body to the top of your head, like water filling a bucket on rainy afternoon.

Choiceless awareness is similar to an open idea space. It takes varying amounts of effort and directs your awareness to whatever arises in your idea space, whether it be the breath, a thought, a sensation, an emotion, a sound, etc. (figure 60-c). In a sense, it's like taking a panoramic lens photograph. To practice this, let attention roam free for 20 seconds, like a dog without a leash. Notice whatever arises naturally.

Selfless awareness is similar to a clopen idea space. It is when you become aware of awareness itself (figure 60-d). It's like being aware that the *idea of the idea space* is itself an idea within your idea space. This practice is objectless and requires minimal effort. It's as if you are aware of taking the photograph itself. To experience this, take a second to note any surrounding sounds. Ask yourself: *Is there a hearer doing the hearing? Or, is there simply the act of hearing?* Take a step back. Turn awareness 180 degrees onto itself. Expand your awareness to the space around you. As Taoist philosopher Wei Wu Wei reveals, "What you are looking for is what is looking . . ."[20] Do you find anyone when you are looking? *No.* Not finding anyone *is* the finding. There is only this spacious awareness.

Of course, there are many techniques other than the spectrum of awareness, like Zen, Buddhism, stoicism, loving-kindness (mettā), yoga, effortless mindfulness, transcendental meditation, Headless Way, etc. In my experience, doing a little bit of each practice helps build a well-rounded idea space. If you treat each mindfulness technique as a tool, then having a diverse toolkit helps you overcome any circumstance.

The main difficulty with grasping mindfulness is it is difficult to measure. For instance, if someone were to look at you, would they know whether you're being mindful, or not? Would they know whether you're focused on the door, or the wall?

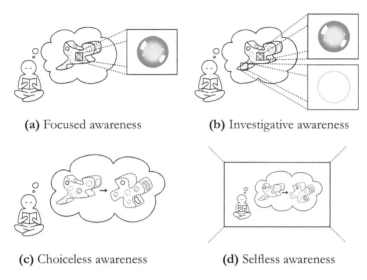

(a) Focused awareness   **(b)** Investigative awareness

(c) Choiceless awareness   **(d)** Selfless awareness

Figure 60. Different types of mindfulness practices.

Furthermore, mindfulness is a difficult experience to share with others. This is natural, as the act of mindfulness sits within your idea space, which has zero measure. So, how do you share something that cannot be seen by others? This creates a situation where people may understand the benefits of mindfulness but have difficulty "getting" mindfulness. This is evident in the most frequent question beginners ask: *How do I know if I'm doing it right?*

The simple answer is mindfulness is like an ON/OFF switch. You're either mindful, or you're not. Focusing on the breath shows you what being mindful looks like. Getting lost in thought shows you what being unmindful looks like. Continuously questioning whether you're being mindful or not is a classic trap. To combat this, we turn to Zen master Seung Sahn Sunim, who said, "Don't check. Just go straight."[21]

While the simple answer may work for some people, it may not work for all. This chapter develops a more explicit definition of mindfulness to clear up what mindfulness is and isn't. Doing so will lift the next veil of illusion on our Path of Awakening by introducing an object, called a *topological singularity*.

A topological singularity is a set of objects whose defining characteristic is that they look like nothing, because they have zero measure, but once uncovered, they unveil their uncountable depth. For instance, your idea space

MINDFULNESS

is a prime example of a topological singularity. It looks like nothing to others, but, clearly, there is an infinity hidden behind the veil of nothingness. An infinity only you can see. Furthermore, as we shall see toward the end of this chapter, mindfulness can be defined as awareness of your idea space in between topological singularities. When you are not mindful, you don't see anything. When you are mindful, you see everything. Other examples of topological singularities include a veil of illusion being lifted, a koan forming in your idea space, the observable universe during the Big Bang, and the Cantor Set (figure 61).

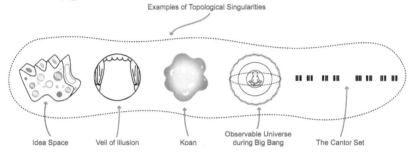

Figure 61. Topological singularities are objects that look like nothing, because they have zero measure, yet are uncountably deep.

That said, it can be difficult to fully understand what a topological singularity is. After all, how can something simultaneously have zero measure and be uncountably deep? Aren't infinity and zero polar opposites? To answer these questions, we'll explore what a topological singularity is at a high level. Then, we'll introduce a more tangible example of a topological singularity in the *Cantor Set*. With this framing, we'll be able to define mindfulness more clearly.

TOPOLOGICAL SINGULARITY

A topological singularity is any object with zero measure and uncountable depth. Since it has zero measure, it looks like nothing, ∅, to an outside observer (figure 62). For example, your idea space is a topological singularity. You can check that your idea space has zero measure simply by closing your eyes and picturing an object, like a book or computer, in your idea space. Can anyone else see the object in your mind? No. You might counter with the fact an fMRI can develop a picture of the brain or new machine learning models can recreate what we see, but the fancy light show and reconstructed images are not a perfect replica of everything you feel, see, think, and perceive.[22] Hopefully, in the future, Elon Musk and Neuralink will find a way to bypass this problem. But, for now, this is where we are.

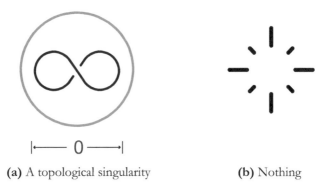

(a) A topological singularity (b) Nothing

Figure 62. A topological singularity, which is uncountably deep and has zero measure, looks like nothing to an outside observer.

The conundrum then becomes: *If you're looking at nothing, are you really looking at nothing, or infinity hidden behind the veil of nothingness* (figure 63)? For instance, look at the space between your eyes and this book. Can you tell whether there is truly nothing there or an infinite vastness hidden from the world? No—the two look the same.

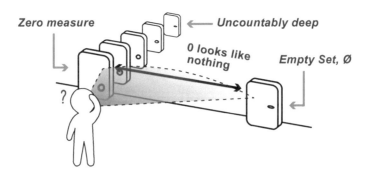

Figure 63. Behind the veil of nothingness can lie an uncountable depth: a topological singularity.

MINDFULNESS

Topological singularities have two other notable properties. First, some can be uncovered, while others cannot. In other words, in certain instances, you can see past the veil of nothingness to uncover the infinity hidden beyond. Other times, no matter how hard you look, you will always see nothing. Second, topological singularities can be nested within one another—you can have a topological singularity within another topological singularity.

For instance, your idea space is a topological singularity that cannot be uncovered by others. Within your idea space lie other topological singularities, like when a veil of illusion is lifted, or when an Unknown Unknown enters your idea space. Before entering your idea space, the Unknown Unknown looks like nothing. However, as soon as the veil is lifted, its uncountable depth shows itself to you. For a specific example, think of the term "idea space." Ten years ago, it looked like nothing in your idea space. Now, you have a better understanding of its uncountable depth.

THE CANTOR SET

To better understand how an object can simultaneously have zero measure, yet be uncountably deep, like your idea space, we need to introduce the magical Cantor Set, which is the most tangible example of a topological singularity. This section is a bit more challenging than other sections, so I've added various elements to provide cognitive ease.

What is the lore of this fairytale beast? *Seeing is easier than understanding the words* (figure 64). We start with the unit interval [0, 1]. Then, we remove the open middle thirds: (1/3, 2/3). This leaves us with the remaining closed intervals [0, 1/3] and [2/3, 3/3].[*] Then, we remove the open middle thirds of the remaining two intervals. Repeat this pattern ad nauseam. As you can see, the more you iterate, the more intervals there are and the smaller those intervals become. I've added Box 1, which color codes several elements for cognitive ease and breaks the Cantor Set down for a couple of iterations, so you can get a better feel for this mystical creation.[†]

---

[*] Please note any number over itself is simply one. Explicitly, $3/3 = 9/9 = 81/81 = \ldots = 1$.

[†] If you are seeing this in black and white, please take a moment to digest the figure caption and Box 1.

94   THE IDEA SPACE

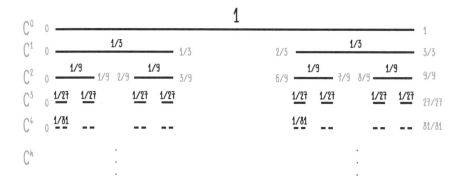

Figure 64. Constructing the Cantor Set. In **blue**, we have the start and end points of each interval. Iterations are noted with a capital "*C*" in red. The measure of each iteration is in **black** above each line segment.

More formally, the Cantor Set, C, is the intersection, or what's in common, between each iteration (figure 65-a). In other words, the Cantor Set consists of the intervals of the last iteration you performed. For instance, if you only performed one iteration, then the Cantor Set would be C^1. If you did 27 iterations, then the Cantor Set would be C^{27}. After 100 iterations, the Cantor Set would be C^{100}. As you keep iterating and zoom out, some say the Cantor Set looks like a veil being lifted (figure 65-b).

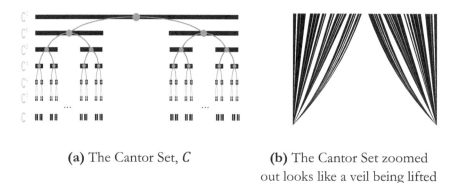

(a) The Cantor Set, C (b) The Cantor Set zoomed
                                    out looks like a veil being lifted

Figure 65. The Cantor Set.

> Box 1
> # Understanding the Cantor Set
>
> Let's break the Cantor Set down for a couple of iterations:
>
> C^0: We start with C^0 which constitutes the unit interval *[0, 1]*. This has a measure of **1**.
>
> C^1: Remove the open middle thirds of the unit interval to obtain C^1. This becomes the union of two intervals *[0, 1/3]* and *[2/3, 3/3]*. The total measure is: $1/3 + 1/3 =$ **2/3**.
>
> C^2: Take away the middle thirds of the remaining two intervals and you're left with C^2. This is four intervals *[0, 1/9]*, *[2/9, 3/9]*, *[6/9, 7/9]*, and *[8/9, 9/9]*. The total measure of these intervals is $1/9 + 1/9 + 1/9 + 1/9 =$ **4/9**.
>
> C^3: You can continue this for C^3 and you get eight intervals:
>
> > *[0/27, 1/27],*
> > *[2/27, 3/27],*
> > *[6/27, 7/27],*
> > *[8/27, 9/27],*
> > *[18/27, 19/27],*
> > *[20/27, 21/27],*
> > *[24/27, 25/27],*
> > *[26/27, 27/27]*
>
> Each of the eight intervals has a measure of 1/27 for a total measure of **8/27**.
>
> C^{27}: You can continue this pattern indefinitely. For example, at C^{27}, you would have 227 intervals, each with a measure of $(1/3)^{27}$, for a total measure of $2^{27} \times (1/3)^{27} =$ **$(2/3)^{27}$**.
>
> **Takeaway: The total measure of the Cantor Set gets smaller and smaller as you iterate.**

The question becomes: *What happens when we keep iterating toward infinity?* At infinity, you are left with an infinite amount of endpoints, which are often dubbed "Cantor Dust" (figure 66). This leads to two other questions: What is the total measure of the Cantor Set as we approach infinite iterations? And, does this set have a countable or uncountable number of points?

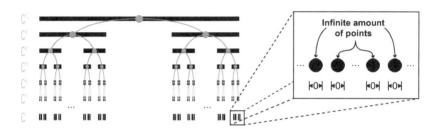

Figure 66. As you iterate toward infinity, the Cantor Set has an infinite amount of points known as "Cantor Dust."

Let's start with the question of measure. It is essential to understand the concept of measure in the Cantor Set differs from our intuition of size or length. The measure of the Cantor Set is the total length of the remaining intervals after each iteration. And, for each iteration, more material is removed, so the total measure of this Frankenstein gets smaller and smaller (see Box 1).

Now, it is true Cantor Dust particles contain measure if we only consider a finite number of iterations. However, when we talk about Cantor Dust, we are referring to the set of points remaining after infinite iterations. At this stage, the remaining points have no measure because they no longer form intervals. They are isolated points that, as a collection, have a total measure of zero.

To illustrate this last point: What is the distance between an arbitrary point, a, and itself? Zero. Now what if you add two points together? Well, zero plus zero is still zero. Even if you add an infinite amount of zeros together, you will still get zero. Ultimately, as we approach infinite iterations of the Cantor Set, we are left with an infinite number of points; and, the total measure of an infinite collection of points is zero. Therefore, the Cantor Set has zero measure.

The fact that iterations can continue indefinitely does not imply the measure of the Cantor Set never reaches zero. In fact, the measure *converges*

MINDFULNESS

to zero as the number of iterations *approaches* infinity. This is a mathematical result that comes from the specific construction of the Cantor Set and the way the Cantor Set's measure decreases with each iteration.

To answer the question around how many points there are, let's bring back our best friends, Joe and Misty. Remember, the Cantor Set is countable if there is a bijection, or a mapping that's one-to-one and onto, between all the points in the Cantor Set and the natural numbers, $\mathbb{N}$ (1, 2, 3, 4, ...). If there is *no bijection* between the two, then the Cantor Set would be uncountable, as Cantor's Diagonal Argument will shortly demonstrate.

After hearing of the last counting disaster, the company behind Deep Rule decided to send our friends an updated version: *Deep Rule 2.0*. The letter said: "Dear life forms, we are sorry to hear Deep Rule did not meet your needs. To make up for this, please find Deep Rule 2.0. We've integrated the latest tech and are confident it will achieve its sole, designed purpose: to count everything."

To see if Deep Rule 2.0 can live up to the hype, Joe and Misty decide to test whether the Cantor Set is countable or uncountable. Misty takes Deep Rule 2.0 out of the gift bag and places it on the ground. She turns the machine on and instructs it to count all the points in the Cantor Set.

A powerful laser shoots out, and within milliseconds Deep Rule 2.0 prints out in the same robotic voice as before, "Counting points of Cantor Set. Must introduce geometric coding. For the first iteration, 0 means go left. 2 means go right. For the second iteration, 00 means left left. 02 means left right. 20 means right left. 22 means right right. For the third iteration, 000 means left left left . . . See screen for further details (figure 67)."

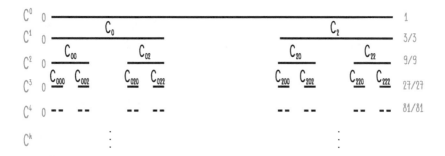

Figure 67. The geometric coding of the Cantor Set: 0 means left, 2 means right.

After Joe and Misty inspect the image, Deep Rule 2.0 continues, "Life forms, please think of this geometric coding as addresses. For instance, the left most interval in the 27th iteration, C^{27}, lives at the following address: $C_{000000000000000000000000000}$. C with 27 zeroes. With infinite iterations, we have an infinite number of points. I have used this coding to successfully label all the points in the Cantor Set. For example, the left most point will have an address of $C_{00...00}$ with infinite zeroes. The right most point will have an address of $C_{22...22}$ with infinite twos. See screen for the full list of infinite addresses in the Cantor Set (figure 68). You will be pleased. I have counted them all as is my sole, designed purpose."

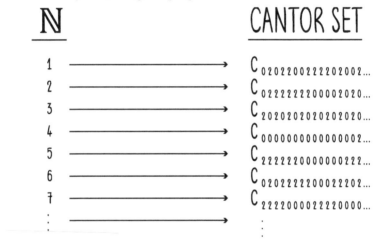

Figure 68. Deep Rule 2.0 counting the infinite list of addresses in The Cantor Set.

Joe looks at Misty and winks. "Deep Rule 2.0," Joe says, "I see you've counted an infinite number of points, or addresses. But I still think you missed one."

Deep Rule 2.0 replies, "Impossible. My calculations are perfect."

Joe points at the screen and says, "What if I picked the point: $C_{2002002...}$? In other words, for each iteration, whenever you go left, I go right, and vice versa. When you pick 0, I pick 2; when you pick 2, I pick 0." Joe gets out a red pen and draws it on the screen for Deep Rule 2.0, eventually showing a red, diagonal line (figure 69).

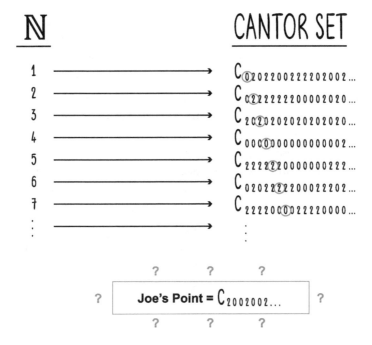

Figure 69. Where is Joe's point in the infinite list of points Deep Rule 2.0 counted?

"Look," Joe says, "my first number is different than the first number you used in your first address. My second number is different than the second number you used in your second address. My third number is different than the third number you used in your third address . . . In other words, my new point is different than every address in your list, because it differs from each of your addresses in at least one position."

Joe asks, "Tell me, Deep Rule 2.0. You said you counted every possible point in the Cantor Set, but where is my point in your infinite list of addresses? In fact, for every point you choose, I can always create a new point that's not in your list by choosing the opposite direction of the point you list. If you choose 2 then I'll choose 0. If you choose 0, then I'll choose 2."

Joe then looks at Misty and says, "These toys suck. I'm going back to using a good ol' fashioned ruler to measure things."

Upon hearing this, Deep Rule 2.0 responds, "My pur-pose . . . is . . . de-stroyed . . . Can-not . . . com-pute . . . Must . . ." and then spontaneously flies in the air and explodes like a beautiful firework. The End.

By employing Cantor's Diagonal Argument, we can conclude the Cantor Set consists of an uncountable number of points. If the Cantor Set had a countable number of points, as Deep Rule 2.0 attempted to demonstrate, there would be a bijection—a one-to-one and onto mapping—between the natural numbers, ℕ, and the infinite points in the Cantor Set. However, Joe successfully revealed there will always be a missing address in Deep Rule 2.0's list. This is because Joe can consistently select a point not included in Deep Rule 2.0's list by choosing the opposite direction of the point at each address provided. The continuous emergence of new points indicates the mapping between the natural numbers, ℕ, and the Cantor Set's infinite points is no longer onto, thereby breaking the bijection. Consequently, with a broken bijection, the infinite points are no longer countable. In simpler terms, for every point or address selected in the Cantor Set, there will always be an unlisted point or address. Thus, the Cantor Set is uncountable.

Having explored the fascinating properties of the Cantor Set, we can now appreciate its resemblance to your idea space as a topological singularity. Just as the Cantor Set is uncountable and has zero measure, so too is your idea space filled with infinite thoughts, emotions, sensations, and perceptions that are inaccessible to others (figure 70). Moreover, just as the Cantor set continuously divides into smaller and smaller subsets, our minds also generate a continuous stream of sub-ideas, reflecting the fractal nature of our thoughts and perceptions. To the world, your idea space may appear as nothing, ∅, but clearly there is something there. Within this place of zero measure lies a vast and unique inner world only you can see. All you have to do is *look*.

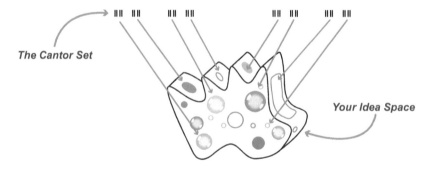

Figure 70. Your idea space is uncountable and has zero measure, just like the Cantor Set.

Moving forward, we'll represent a topological singularity using the symbol of the uncountable points of the Cantor Set condensed into a single point (figure 71).* This compact and powerful symbol captures the pure essence of a topological singularity, effectively conveying the paradoxical nature of an object that is both uncountable and has zero measure. A true Zen-like oxymoron.

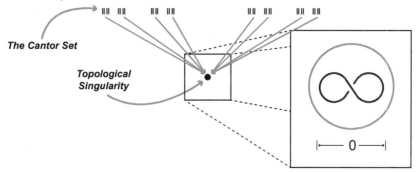

Figure 71. A topological singularity is represented by the Cantor Set crushed into a point.

MINDFULNESS

Where do thoughts come from? Where do thoughts go once noticed? While these seem like trivial questions, there is no clear answer to them. Thoughts, emotions, sensations, and perceptions simply arise out of what feels like thin air, or nothingness. Then, as soon as these elementary ideas arise, you can become mindful of them and their uncountable depth becomes clear. However, the impermanent nature of our world soon causes them to vanish. But where do they go once noticed? No one knows. They go back to looking like nothing.

When you are not mindful, or lost in your wandering mind, awareness of your idea space looks like nothing, ∅. You are not *looking* at anything. Then, the instant you become mindful, by focusing on the breath or feeling the sensations in your hand, the whole world unveils itself—awareness of your idea space shows its uncountable depth. Thus, the topological

---

* The Cantor Set is totally disconnected: no two points touch, or overlap. Crushing them to a point provides the trivial solution to connecting the uncountable points of the Cantor Set.

singularity is unveiled. The switch is turned ON.

Unless you are an *arahant* (Pali), or fully awakened individual, mindfulness comes in moments. As an Ancient once said, "Short moments, many times."[23] It is common to be mindful in one moment, then get sucked back into your wandering mind. In other words, you return to not *looking* and your idea space once again looks like nothing. You return to the topological singularity. The switch is turned OFF. Thus, mindfulness is awareness of your idea space in between topological singularities (figure 72).

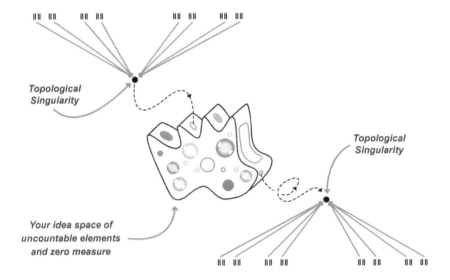

Figure 72. Mindfulness is awareness of your idea space in between topological singularities.

The beauty of mindfulness is the topological singularity can always be uncovered. You can always come back to looking. The easiest way is by noting the arising and passing of your experience or seeing the impermanent reality of the world. Usually, a single word to note the experience works best. Thought. Emotion. Sensation. Perception. Change.

Another way to become mindful is to do a mindful pause, or return to beginner's mind with an empty idea space, by performing a S.T.O.P.P. There, for an instant, you will see the nonduality nature of reality since nothing, ∅, is clopen.

That said, it is a common pitfall to linger in emptiness. Indeed, many meditators find it challenging to focus on "nothing," as the mind tends to wander and latch onto various thoughts and sensations. However, the goal of meditation is not to force yourself to think of nothing, but rather to cultivate an awareness and acceptance of the present moment without judgment. As Dzogchen Zen master Nyoshul Khen Rinpoche said:

> Those who believe in things can be helped through various kinds of practice, through skillful means—but those who fall into the abyss of emptiness find it almost impossible to re-emerge, since there seems to be no handholds, no steps, no gradual progression, nothing to do.[24]

It is essential to strike a balance between recognizing emptiness and not becoming trapped in it. In this context, lingering in emptiness refers to the risk of becoming excessively attached to the idea of emptiness and losing sight of the broader practice of mindfulness and presence. Or, as an Ancient once said:

> Better you should give rise to a view of existence as big as Mt. Sumeru, than that you produce a view of nothingness as small as a mustard seed.[25]

Emptiness is only part of the picture, not the whole story. If you are lost in emptiness, remember the complement of nothing is everything. In order to see nothing, you must see everything.

Personally, I've found being at awe with world, as if I was a child seeing it for the first time, is a great trick to remain mindful. As Albert Einstein said, "There are only two ways to live your life. One is as though nothing is a miracle. The other is as though everything is a miracle." This involves dropping any preconceived notion of what you think the world is and simply taking it as it is: hearing, listening, seeing, thinking. The world is coming toward you; you are not going toward it.

Of course, it is not practical to live in a constant state of awe. Instead, I use the short moment of awe to see the beauty of the world, which helps me find beauty in more ordinary tasks, like breathing, sensing, cleaning, walking, being sick, etc. As Confucius said, "Everything has beauty, but not everyone can see."

Maintaining mindfulness, even when experiencing awe or appreciating the beauty of ordinary tasks, can be a challenging task. Amid these challenges, it is natural to sometimes feel disheartened or frustrated. However, a staple of Zen is good humor. As Lin Yutang writes in *The Importance of Living*, "Reality + Dreams + Humor = Wisdom."[26] Or, as a Zen master once said, "Nothing is left to you at this moment but to have a good laugh."[27] It is important to have a good sense of humor about your own mental foibles.

So, *how do I know if I'm practicing mindfulness correctly?* Stop for a second and take a few deep breaths. Simply focus all your attention on the rise and fall of the breath. Where do you feel it the most? What does shifting your awareness to that point feel like?

It is trivial to see when we are mindful; but, for some reason, we constantly second-guess ourselves. An Ancient once said, "It is easy to see in the phenomenon: if you try to figure it out in your mind, you will lose contact with it."[28] Do not try to seek mindfulness out. Let the experience come to you.

THE KOAN EXPERIENCE

Overall, understanding mindfulness as awareness in between topological singularities lifts the next veil of illusion on our Path of Awakening (figure 73). Prior, it was unfathomable to think of a coexistence between zero and infinity. Now, we see the two live in clopen harmony whenever we enter a mindful moment. Simply turn the switch ON. Or, as professor Timothy Leary said, "Turn on. Tune in. Drop [the Self] out."

Up to this point, we've developed a robust model for looking at the mind objectively, called your idea space. Your idea space is unique to you, uncountable, and has zero measure. Within your idea space lie various thoughts, emotions, sensations, and perceptions that can be clopen. The task is then on you to apply mindfulness into your day-to-day, so you can appropriately handle pleasant and unpleasant experiences. In the remaining chapters, we'll see how your idea space relates to the universe at large.

Mindfulness experiences come in many different shapes and sizes. Some are straightforward, like focusing on the breath, feeling a touch

MINDFULNESS

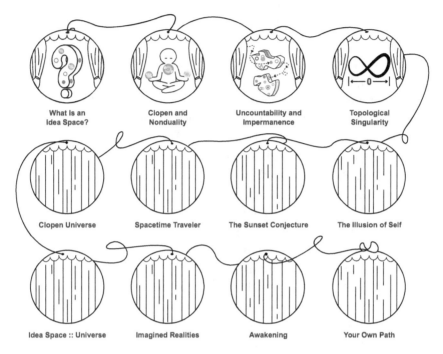

Figure 73. A topological singularity, an object that is simultaneously uncountable, yet has zero measure, lifts the fourth veil of illusion on our Path of Awakening.

sensation, hearing a sound, noticing a thought, etc. Others are more abstract, like koans or realizing everything we see is in the past. As we'll see later in the book, light takes time to travel from point A to point B, even when traveling at 186,000 miles per second. So, for example, when you're looking at the sun, you're seeing how it was eight minutes ago, because it took eight minutes for light from the sun to reach you.

Although abstract, koans, which are paradoxical statements or questions used in Zen Buddhism to provoke deeper insight and understanding of the nature of reality, can be quite liberating. They capture an experience that transcends space, time, and perception. A koan seizes the identity of a particular idea space, or all the thoughts, emotions, sensations, and perceptions of a particular moment in spacetime, that would otherwise be ineffable. Koans invite us back to a primordial experience by delivering a profound awakening. The moment lasts only for an instant, but its remnants last for an eternity.

Not all koans will hit. It depends on the person and where they are in life. That said, the koans that do hit, hit like a brick wall. At first, they look like nothing. But, as soon as you uncover the koan, the whole world shows itself to you, leaving behind a treasured principle. The koan experience is in itself a topological singularity.

We've already explored a few koans questioning the nature of buddha, or an enlightened mind. These koans revealed that buddha nature cannot be easily defined or grasped, as buddha nature is simultaneously the mind and not the mind. Additionally, other koans use unconventional and surprising references to describe buddha nature.

For example, one koan refers to buddha nature as Mu, which means "no-thing," or the absence of any specific concept. Another koan describes buddha nature as "three pounds of hemp," a seemingly ordinary object, emphasizing the inherent ordinariness of enlightenment.[29] Another koan refers to buddha nature as a "dry shit-stick," which was an ancient form of anal cleansing.[30] This provocative, yet humorous image challenges our preconceptions and encourages us to see enlightenment in even the most mundane or unclean aspects of life.

More koans are made almost every day in your life! It's the aha moment after coming across something that perfectly captures what you've been thinking or feeling. A small moment of enlightenment or awakening. This nonverbal experience can come in the form of a word, phrase, song, image, body language, joke, sight, sound, veil of illusion, etc.

For instance, imagine you are walking on a sunny day, listening to music, and feeling a little overwhelmed with the stresses of life. Suddenly, a new song comes on that beautifully capture your mood, thoughts, and emotions. In this moment, you feel understood and may even laugh a little. Other times, you'll be listening to a podcast and the guest will say a profound quote that strikes a chord within you. When you come across a koan that resonates with you, write it down! Future you will thank you for the principles koans help you build.

To experience the wonders of the abstract koan world, I would like to leave you on a classic koan. Take a few deep breaths to settle in . . . Sit in an alert, yet comfortable position . . . Let go of any preconceived notions of the world . . . Pay particular attention to what arises in your idea space when reading the koan . . . Remember, the world is coming toward you;

you are not going toward it . . . After reading the koan, take a moment to sit with it, and reflect on the experience . . . Ready?

What is the sound of one hand clapping?
For the single hand doesn't sound without reason.

Chapter 5

THE ILLUSION OF SELF

"The true Self is Non-Self."
- Alan Watts

Take a moment to perform a S.T.O.P.P. Notice what arises. What thoughts are present? What emotions do those thoughts lead to? Are they pleasant, unpleasant, or neutral? What sensations in the body are linked to those emotions? Then, expand your awareness. Let your body act as a single cloud of sensations floating in space. In this moment, can you put a smile on your face? Take a few seconds to close your eyes and take a few, deep breaths.

Non-Self characterizes the fact that "I," your name and identity, is simply another appearance in your idea space. It is the amalgamation of your thoughts, emotions, sensations, perceptions, and consciousness. In other words, "I" is merely one layer of your Self. This is the veil of illusion commonly known as the *Illusion of Self*.

To test this hypothesis, let's do a simple experiment called the *Layers of the Onion* from the Headless Way, a meditation technique developed by Douglas Harding. The technique aims to answer the age-old koan:

What is my original face, before my parents were born?

As with any koan, do not try to logic an answer. Turn off your prefrontal cortex, or CEO of the brain. Simply notice what sensations arise.

To start, sit in a comfortable, yet alert position. Imagine asking someone standing ten meters away, "What am I?" They'd probably answer: a person.

Now, imagine that person is looking at you through a microscope and you ask them, "What am I?" They'd probably answer: a mixture of cells. If they view you with an electron scanning microscope, they'd probably answer: molecules. As close as they can get, they see nothing. They cannot see who you are at zero measure. They cannot see your idea space.

Similarly, an observer could zoom out and you could ask them the same question: "What am I?" As they zoom out, they'd probably answer: a city, a country, a solar system, a galaxy, and a universe.

So, what are you? A universe? A solar system? A city? Your name? Cells? Molecules? A mixture of all these things? Where do you start and where do you end? The words of physicist Richard Feynman capture the sentiment well:

> What is a chair? The atoms are evaporating from it from time to time—not many atoms, but a few—dirt falls on it and gets dissolved in the paint; so to define a chair precisely, to say exactly which atoms are chair, and which atoms are air, or which atoms are dirt, or which atoms are paint that belongs to the chair is impossible. To define . . . a single object is impossible, because there are not any single, left-alone objects in the world—every object is a mixture of a lot of things, so we can deal with it only as a series of approximations and idealizations . . .[31]

In this spirit, "I," your name, is an idealization others, and sometimes even you, use to approximate who you are. In reality, your Non-Self is the amalgamation of all your fractal layers (figure 74). These layers extend to the outer edge of the cosmos as everyone lives at the center of their own observable universe. In other words, everyone experiences their own *Singularity Sunset*, as we'll see in the next chapter through the *Sunset Conjecture*.

There was not always an "I" for us growing up. As infants, we had no association with the name our parents gave us. There was only the world. If you have ever owned a pet, then you will be familiar with this fact. It takes time for a dog to understand its name. Once the animal realizes its name, it responds to its identity. The dilemma is the dog's name, or identity, is projected onto it by others, not itself. If others were not there, then the dog would have no name. Similarly, understanding your Non-Self requires you to take a step back. Who are you prior to your name? What is your original face, before your parents were born? Who is doing the seeing? Who is doing the hearing? Who is doing the thinking? Remember, not finding anyone is the finding. What happens if we drop everything we know and attune to what is actually given?

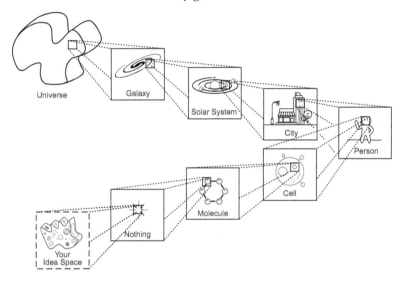

Figure 74. Zooming in and zooming out shows the different layers of your Non-Self.

At first, it may be daunting to realize there is no "I." Over time, this becomes a liberating experience. There is a lot associated with "I," like "Who am *I* supposed to be?," "What am *I* supposed to do?," "*I* have so many problems," etc. Our identity forces us to cling: "*I* like this," "*I* don't like that," etc. We attach ourselves to an impermanent identity that is in constant state of change. When we cling to that identity, we are liable to suffer. By realizing the *self* is simply another appearance in consciousness, we can let go of what the *self* clings to and achieve *anatta* (Pali), or selflessness.[32] Thus, we can reduce the amount we suffer. In the words of the Buddha's essential teaching, "Nothing whatsoever is to be clung to as I or mine."[33]

The best way to see your Non-Self is to focus on the impermanent nature of the world. You are not the person you were five years ago, and you are not the person you will be five years from now. The "I" from then is different from the "I" of today. On a more granular level, "I" is always changing from moment to moment as the world is impermanent. You always have the opportunity to reinvent yourself, as we shall see more clearly in the next chapter . . .

To more closely see how "I" is merely one layer of yourself, we shall dive into the world of fractals and answer a deceptively easy question:

How long is the coast of Great Britain?

The goal of this problem is to demonstrate more rigorously that what you are changes with range. In other words, depending on the size of your measuring stick, you'll get a different total length than the next person. This includes how others measure you. Then, we'll dive back into the Headless Way so you can truly experience your Non-Self.

THE BEAUTY OF FRACTALS

Overall, fractals are the key to unlocking your idea space. If mindfulness allows for the expansion of open awareness, then fractals make us realize "I" does not tell the whole story. As Joseph Goldstein says, "The 'I' in mind is extra."

So, *what are fractals?* At a high level, fractals relate to the study of roughness. In other words, the study of fractals is the study of how objects change the more you zoom in and out.* One of the key tenets to many, but not all, fractals is their *self-similarity*. This means there's a specific pattern repeated at various levels of the object. For instance, the Cantor Set is a fractal, as you are continuously taking out the open middle thirds of the preceding intervals. Overall, fractal shapes can be found in nature, art, mathematics, and many other places as Benoit Mandelbrot, the father of

---

* A common definition of fractals is any set for which the Hausdorff Besicovitch dimension strictly exceeds the topological dimension. In simple terms, this dimension tells us how much space a fractal occupies in relation to its size. For example, some fractals have a dimension between 1 and 2, which means they have a rough or jagged boundary that fills more space than a regular line. Other fractals have a dimension greater than 2, which means they have a highly crumpled or branching structure that fills more space than a flat surface.

fractals, demonstrates in The Fractal Geometry of Nature (figure 75).[34] Isn't life pretty?

Figure 75. Fractals found in nature: fern (left) and Romanesco Broccoli (right).

Roughness, or how objects look at different scales, is important to study, because our perception of how an object looks on its surface may not reflect its true structure. For example, a seemingly smooth object, like a piece of paper (figure 76-a) may actually have hills, valleys, and fibers when viewed at a closer range (figure 76-b). Conversely, when viewed from a distance, it may appear as a mere dot (figure 76-c). At all ranges, it is still paper; but, depending on your range, the appearance of it changes. This is similar to the Layers of the Onion experiment, where our perception of ourselves changes depending on our level of observation.

(a) Paper looked at eye level

(b) Paper looked at under a microscope

(c) Paper looked at far away

Figure 76. Paper is still paper. But, paper looks very different depending on its range.

Thus, roughness and especially fractals remind us we are constantly fooled by trying to put things into arbitrary groups, like smooth or rough.* If you decide to put an object into a specific group, then that object will likely have an uncountable amount of subgroups within it.

For instance, imagine you have an object, like the chair Feynman talked about. You draw a circle around it, calling this circle a "group." Next, you draw a slightly smaller circle within the first one and call it a new group. You continue this process infinitely, drawing an uncountable number of smaller and smaller circles around the object, labeling each circle as a different "group." However, for every circle you draw, it's always possible to choose another circle you have not drawn yet, since the size of the circle is based on space, which is uncountable (see Chapter 3). This means there are an uncountable number of "groups" within your original grouping of the object. Moreover, your initial group is likely a subgroup of an infinite number of other groups. This makes your first grouping completely arbitrary—it might still be useful, but it's important to recognize its arbitrary nature!

For a concrete example, "I," your name, is a useful grouping. There are an infinite number of subgroups within you: your quarks, atoms, molecules, cells, organs, and other biological systems. Then, you belong to an infinite number of larger groups: your city, country, continent, planet, solar system, galaxy, galactic cluster, and observable universe. At one point, we can identify with our country. At another point, we identify with an aching tooth. We all expand and contrast like this all the time. By understanding that our identity consists of countless layers and is constantly changing, we can let our curiosity drive us and embrace the fluidity of our experience to cultivate a more open and adaptable perspective on life.

Plus, at the end of the day, fractals look pretty. It's nice to wake up and smell the roses every once in a while.

HOW LONG IS THE COAST OF GREAT BRITAIN?

Answering the question, "How long is the coast of Great Britain?", shows us more clearly that what we are changes with range. It is an illustrative example that demonstrates there is no *true length* to the coast of Great

---

* Plus, as we unveiled in Chapter 2, grouping is the most fundamental aspect of the human experience—even prior to counting or making measurements.

THE ILLUSION OF SELF

Britain. In the same vein, the concept of "I" is inherently mutable, as our identity constantly transforms depending on the viewpoints of others.

The main reason there is no true length to the coast of Britain is because the total length of the coast depends on your measuring stick. For example, you could choose to measure the coastline with a meter stick, yard stick, or a 12-inch ruler. You may even feel audacious enough and try with a "stone throw" (personally, I'd love to see this).

Depending on your measurement stick, you'll get a completely different answer than the next person. For instance, when we use a measuring stick of 200 km, the coast of Britain is approximately 2,400 km (figure 77). When the measuring stick gets to 50 km, the total length of the coast now goes to 3,400 km. According to the Central Intelligence Agency (CIA), the length of the coast is around 12,429 km.[35] So, which is it? What happens if you use the smallest measuring stick possible?

(a) Measuring stick: 200 km
Total length: 2,400 km

(b) Measuring stick: 50 km
Total length: 3,400 km

Figure 77. How long is the coast of Britain? Well, it depends on your measuring stick.

To answer these questions, we need to introduce a new fractal friend: the *Koch Curve* (figure 78). Reiterating an earlier point, seeing the picture is easier than understanding the words. To construct this legendary creature, we start like the Cantor Set with E_0, or the line interval [0, 1]. This is our *initiator*. Now, at each iteration, wherever we see our initiator, in this case a straight line, we'll replace it with E_1, our *generator*. And, like the Cantor Set, we'll reduce the size of the lengths by a factor of 1/3. Box 2 breaks the Koch Curve down for a couple of iterations for cognitive ease.

As you can see, the more we iterate, the more intervals, or *measuring sticks*, there are. Furthermore, the length of each measuring stick goes down, while the total length of the Koch Curve goes up. Now comes the fun part. What happens if we keep iterating toward infinity? Will there be an uncountable or countable number of measuring sticks? What is the length of each stick? And what is the total measure of this fabled beast?

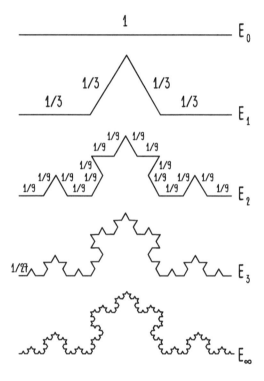

Figure 78. Guide to the Koch Curve. Keep the generator going wherever the initiator is found.

Using a similar geometric coding as the Cantor Set, you can show there are an uncountable number of measuring sticks, each getting smaller and smaller, as we iterate toward infinity. Although each measuring stick eventually reaches zero measure, the total measure of the Koch Curve paradoxically shoots up to infinity.

> Box 2
> # Understanding the Koch Curve
>
> Let's break the Koch Curve down for a couple of iterations:
>
> E_0: We start with E_0, which constitutes the unit interval *[0,1]*. This has a measure of **1**.
>
> E_1: Replace the straight line with the generator to obtain E_1. This consists of four intervals, or *measuring sticks*, whose total measure is: **1/3 + 1/3 + 1/3 + 1/3 = 4/3**.
>
> E_2: Reduce the generator by 1/3 and replace each straight line with the generator. This consists of 16 measuring sticks, each with a length of 1/9. So, the total measure becomes: **16 × 1/9 = 16/9**.
>
> E_3: You can continue this for E_3 and you get 64 measuring sticks, each with a length of **1/27**. The total measure is now: **64 × 1/27 = 64/27**.
>
> E_{27}: You can continue this pattern indefinitely. For example, at E_{27}, you would have 4^{27} measuring sticks, each with a measure of $(1/3)^{27}$. The total measure then becomes: $4^{27} \times (1/3)^{27} = (4/3)^{27}$.
>
> **Takeaway: The total measure of the Koch Curve gets larger and larger as you iterate; and the measure of each measuring stick gets smaller and smaller.**

This creates a captivating juxtaposition: Although the total measure of the Koch Curve approaches infinity, to an outside observer who is unable to perceive the infinite intricacies of the curve, it may appear as if it consists of an uncountable number of infinitely small, insignificant points, each with zero measure. Thus, the Koch Curve, in all its grandeur, could paradoxically seem like nothing, ∅, to the uninitiated observer.

Think about that. To an outside observer, an object with uncountable elements and *infinite measure* can look just like an object with uncountable elements and *zero measure*: nothing.* As Robert Pirsig writes in *Zen and the Art of Motorcycle Maintenance*, "Some things you miss because they're so tiny you overlook them. But some things you don't see because they're so huge."[36]

Look at the space in between your eyes and this book once more. Is nothing there? Is something uncountable with zero measure there? Or is something uncountable with infinite measure there? No one knows because it all looks like nothing (figure 79). Welcome to the wacky world of what it means to be human.

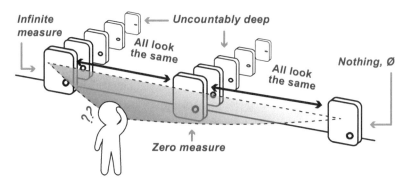

Figure 79. If you're looking at nothing, ∅, are you looking at something that has **(a)** infinite measure, **(b)** or zero measure, or **(c)** is it actually nothing?

Overall, the Koch Curve demonstrates the coast of Great Britain will eventually have an infinite length if you decide to use a smaller and smaller measuring stick. So, there is no definitive, or "true," length to the coast of Britain. It all depends on your measuring stick.

This answer may not satisfy your intellectual taste buds, so we turn to physics. In physics, the smallest supposed measuring stick is called *Planck Length*. Its value is approximately 1.6×10^{-35} meters, or, writing it out completely, 0.000000000000000000000000000000000016 meters (that's 34 decimal places of zeroes)! You can then compare Planck Length to the *self-similarity dimension* of the coast of Britain and the Koch Curve to

---

* Fun fact: there are an *uncountable* number of (a) sets with infinite measure and uncountable elements and (b) sets with zero measure and uncountable elements. For every one of these sets you find, I can always find one you didn't find.

approximate the total length of the coast to be around 500,000 km.* Quite different than the 12,429 km found on the CIA website.

THE SCIENCE OF THE FIRST PERSON

The purpose of that demonstration was to show you that what you are changes with range. Using a certain measuring stick, the coast of Britain had one total length. Using another, it had a completely different length. No matter which stick you used, it was still the coast of Britain. Similarly, you change depending on people's perspective of you. As close as others can get to you, they see nothing. They cannot see your center, or your idea space. They cannot peel away the final layer. At this point, the science of objects, the science of things we can measure, reaches its limits.

As always, absence of evidence is not evidence of absence. Simply because it *looks* like there is nothing there, does not mean there *is* nothing there. You are here at center. You are at this place of zero measure. Clearly, there is something here. An infinite vastness. A unique, layered perspective of the universe. This is the science of the first person—the science of things with zero measure.

To appropriately study the science of the first person, we need to develop a proper set of tools analogous to the tools in the science of objects. For things we can measure, we use rulers, sensors, and semiconductors to get a better sense of reality. For the science of the first person, we have to turn to different meditation techniques, like Zen, Stoicism, or even basic mindfulness practices such as breathwork and body scans.

In both sciences, the key is to test ideas for yourself and not take anything at face value. For example, if someone tells you something is ten meters, then take a ruler and measure it yourself. If someone tells you practicing gratitude can improve your overall well-being, then try keeping a gratitude journal to see if it brings positive changes to your life. Of course, you do not need to constantly engage in this practice, but periodically incorporating a skeptic's eye is essential to awakening. As the Buddha once said, "Believe nothing, no matter where you read it, or who said it, no matter if I have said it, unless it agrees with your own reason and your own common sense."

---

* See supplemental material on www.TheIdeaSpace.io for the derivation of this number.

A unique, investigative tool for the science of the first person is called the *Headless Way*, by Douglas Harding. To experience your headless nature, we'll explore a classic experiment, called the *Pointing Experiment*. The goal of this experiment is show you the dividing line between the science of objects and the science of the first person. It shows "I" is a mere tool we use to approximate the first-person experience. Therefore, we should not cling to "I" as the authoritative definition of our true Self, as that will only lead to suffering.

But before we start, a few words of introduction. The headless experience revolves around imagining you have no head; and, in place of having a head, simply seeing the world appear where your head should be. Not having a head does not literally mean being decapitated. Instead, it is a somewhat "childish" technique that invites the skeptic to look at life from the first-person point of view, as if you are your own character in a first-person video game (figure 80).[37]

Figure 80. Your headless nature as you read this book.

Do not struggle with the following experiment. It is not a matter of going deep and providing some unworldly experience. Headlessness lies right on the surface of consciousness and can be glimpsed in a moment. Pay particular attention to how the world appears in the first instant, before thoughts intervene. The resulting glimpse of the first-person experience will only last a moment, so simply repeat this glimpse, again and again, in as relaxed a way as possible.[38] The key to the experiment is to be an honest reporter, describing solely what you observe and nothing else.[39]

To start, take a slow, deep breath. Let go of any preconceived notions of the world, and simply attune to the objective sights given. Keep breathing. Stop for a moment and pick an object in front of you. Any object will do. *Physically point* to the object. Do not stop pointing until I tell you to do so. Take a moment to describe that object for me. What is its shape? What is its size? What is its color? What is its smell? Does it have any defining features?

Now, with the same finger, point to another object. Physically point to it and keep pointing. What is that object's shape, size, color, and smell? How is that object different from the first? How is it the same? Take a moment to answer these questions.

Now, take your pointed finger and physically point back at where others see your face. In your experience, what are you pointing to? Can you describe what you're pointing to? How big is it? What color is it? Is there anything like it to compare it to? As an honest reporter, can you truly see what are you pointing to? You can stop pointing now.

Douglas Harding recalls the experience of the Pointing Experiment well in his book, *On Having No Head*:

> What I found was khaki trouser legs terminating downward in a pair of brown shoes, khaki sleeves terminating sideways in a pair of pink hands, and a khaki shirtfront terminating upward in—absolutely nothing whatever (figure 81)! Certainly not a head.

Figure 81. A representation of what Douglas Harding saw.

It took me no time at all to notice that this nothing, this hole where a head should have been was no ordinary vacancy, no mere nothing. On the contrary, it was very much occupied. It was a vast emptiness vastly filled . . . A nothing that found room for everything . . . I had lost a head and gained a world.

Here it was, this superb scene, brightly shining in the clear air, alone and unsupported, mysteriously suspended in the void, and utterly free of 'me' . . . Lighter than air, clearer than glass, altogether released from myself, 'I' was nowhere around.

It was the revelation, at long last, of the perfectly obvious . . . It was a ceasing to ignore something which (since early childhood at any rate) I had always been too busy or too clever or too scared to see. It was naked, uncritical attention to what had all along been staring me in the face—my utter facelessness.

In short, it was all perfectly simple and plain and straightforward, beyond argument, thought, and words. There arose no questions, no reference beyond the experience itself, but only peace and a quiet joy, and the sensation of having dropped an intolerable burden.[40]

To summarize his experience, Douglas Harding starts by realizing the dividing line between the science of objects and the science of the first person. He can see his khakis, shoes, sleeves, and hands (science of objects), but he sees *nothing* as he keeps going up. No head in sight. Of course, there is not "nothing" there, but instead "a vast emptiness vastly filled." In other words, he sees his uncountable idea space hidden from the outside world (science of the first person). From there, he realizes "I," his name and identity, is nowhere to be found and there is simply what is given—the present moment. He thus drops the "intolerable burden" of "I" and achieves an experience filled with peace and joy.

Overall, "not having a head" is simply another way of viewing your idea space objectively—devoid of "I." It shows the science of the first person and the science of objects can live together in clopen harmony. Namely, someone else's experience of me is very different than my experience of me. Someone else says I have a head. I say I don't. To you, *there*, I do have a head. But, for me, *here*, I do not have a head. Both viewpoints are valid. They are views from different ranges. Growing up, we are told the only valid view of ourselves is from *there*, the science of objects. But the view of myself, from *here*, the science of the first person, is equally as valid.

The experiment demonstrates a key difference between the science of objects and the science of first person. On one hand, in the science of objects, the comparison of sizes and shapes is possible due to the presence of multiple objects, leading to the *relativity of size*. This notion mirrors the observation we made in the Great Britain example: Different measuring sticks yield different total lengths, emphasizing the variability influenced by unique perspectives and ranges.

On the other hand, when looking at the science of the first person, we encounter a singular and indivisible field of vision. Determining the size of this field becomes an enigma, as there exists no other comparable entity against which it can be measured. The first-person experience stands apart, existing at zero measure, concealed from the purview of the science of objects. Consequently, sharing experiences that unfold within the first person realm becomes challenging, as they manifest with remarkable individual variation.

A common objection to the Headless Way is: "I can't see my head, but I can touch it." Can you touch your head? Investigate this. Take your hand in front of you so you can see it. Now, move it toward you, and gently place it on the top of your head. What do you notice?

In my experience, as the hand approaches the top of the head, it disappears. Then, touch sensations appear in awareness. I can imagine what my hand on the top of my head would look like, but I do not see my head nor fingers. The image I create is simply a mental construct, a thought. It is crucial to differentiate between the mental image we construct and the actual experience given to us, which might be different from what we expect. It is crucial to be an honest reporter.

Of course, for someone *there*, they would see my hand touch my head. But, from zero distance, *here*, there is no head here. There is simply space for sensations, thoughts, emotions, and perceptions. What happens when we put aside what we know and attend to what is actually given?

Another objection is: "What about when we look at ourselves in the mirror?" The answer here comes down to the fact everything you see is in the past as it takes time for light to travel from point A to point B, even when traveling at 186,000 miles per second. So, the object you see in the mirror is simply an image of your former Self, albeit a relatively recent image.

The image in the mirror is not your true Self, nor is the singular reflection of your idea space. As Alan Watts said, "The true Self is Non-Self." In other words, your true Self is a combination of both these Selves—the combination of the science of objects and the science of the first person.

NON-SELF

Overall, your Non-Self is the unification of the science of objects, things we can measure, and the science of the first person, things with zero measure. It involves understanding that all the layers of your observable universe are unique to you, from the almost unrecognizable idea space to the outer edge of the observable cosmos: the Big Bang. This realization dissolves the attachment to the concept of "I" we cling to; and, thus, we can lift the next veil of illusion on our Path of Awakening: the Illusion of Self (figure 82). Prior, there was an "I," or a name. Now, there is simply this open, spacious awareness.

Figure 82. The Illusion of Self lifts the next veil on the Path of Awakening.

THE ILLUSION OF SELF

After building a strong, objective framework for looking at the mind in the previous chapters, we shifted our attention to start seeing how the mind relates to the rest of the universe in this chapter. We did this by looking at fractals, which showed you it is impossible to pinpoint exactly where "you" start and where "you" end. In reality, your Non-Self consists of all your unique fractal layers. To see this more clearly, we explored a concrete example asking, "How long is the coast of Great Britain?" Here we were able to clearly see that what you are changes with range.

We then took a deeper dive to investigate the science of the first person through the Headless Way. As we engage in this thought process, we are able to gain a deeper understanding of our own inner landscape and the boundaries between our individual experience and the external world.

The exploration of your own headless nature seeks to answer the profound question, "What is your original face before even your parents were born?" The answer lies in a nonverbal experience that reveals the dividing line between the science of objects and the science of the first person. This realization helps us understand that "I" is merely a tool we use to communicate our first-person experience through the vocabulary of the science of objects. Ultimately, the Headless Way invites us to embrace the unique, individual nature of our experiences, while recognizing the perspectives of ourselves and others can coexist in harmony. By doing so, we can better appreciate the richness and depth of our own existence.

By embracing this nondual perspective, we can cultivate a sense of unity with our surroundings and foster a greater sense of compassion and empathy for others by realizing everyone lives in their own headless world. Moreover, this shift in perspective can lead to profound personal growth, as we learn to let go of our attachments to ego and identity, and instead embrace the interconnectedness of all things. By recognizing that our subjective experience is only one aspect of the vast, interconnected tapestry of existence, we can begin to develop a more open and flexible outlook on life, free from the constraints of rigid self-definition and self-identification. This newfound openness allows us to welcome and appreciate new experiences we might have previously dismissed or overlooked, such as trying a unique cuisine, exploring a different culture, or engaging in an unfamiliar hobby that expands our lives.

In short, your headless nature brings out of hiding your true Self. It is a glimpse into *anatta*, or selflessness. The beauty is you can always return to your headless nature at any point. All you have to do is *look* out of your single, undivided field of vision.

The exploration of the science of the first person through practices like the Headless Way, mindfulness, and meditation can offer invaluable insights into the nature of our own consciousness and its relationship with the world around us. As we continue to delve deeper into this realm of inquiry, we may discover a newfound sense of freedom, joy, and inner peace that transcends the limitations of our everyday perceptions. Ultimately, this transformative process empowers us to live a more authentic, compassionate, and fulfilling life, enriched by a deepened awareness of the boundless beauty and complexity of existence.

Chapter 6
THE SUNSET CONJECTURE

"Our genius is to understand and stand beneath the set of stars present at our birth. And from that place, seek the hidden single star, over the night horizon, we did not know we were following."
- David Whyte

Breathe in . . . Breath out . . . Take a moment for yourself. With everything happening in the world, sometimes it's best to simply sit and focus on the here and now. What is present in this moment? Any sights? Sounds? Sensations? Emotions? Are there any fictitious stories in your mind you're playing on repeat? Simply notice what is present, as you would the breath.

The *Sunset Conjecture* is the bridge between the science of the first person (things with zero measure) and the science of objects (things we can measure). It is best illustrated through a short story:

> Picture yourself on the beach. You had a fantastic day of doing nothing but reading, drinking piña coladas, and talking with friends. At the end of the day, you decide to walk on the beach to watch the sunset. You say to your friends, "I must be special. It seems the sun's golden rays are reflecting right off the water directly toward me." Your friend responds, "No, you idiot. The golden rays are pointing directly toward me." Clearly, you are both right, the situation is clopen (figure 83).*

---

* In this context, clopen signifies two seemingly opposite things being simultaneously true, like open and closed.

Figure 83. The sun sets on two people simultaneously.

This phenomena can best be explained by the Sunset Conjecture, which is two-fold: (a) you live at the center of your own observable universe, and (b) at the center lies your idea space of uncountable depth and zero measure. Since you are at the center of your own observable universe, light from any object hits you uniquely. For instance, like the sunset, the moon's silver rays hit you differently than anyone else. So too do the gravitational effects of the giant black hole at the center of the Milky Way.

All in all, everyone's reality is different, as everyone experiences their own perspective of the universe. In other words, you can pick any object in the universe, and it will have a unique impact on you. As we shall see, this includes the edge of the observable universe, the Big Bang. In this instance, everyone experiences their own *Singularity Sunset*.

To better understand how the human experience is unique to all, we'll take a deeper look at cosmology, or the origin and development of the universe. In doing so, we'll lift the next veil of illusion on our Path of Awakening through the Sunset Conjecture.

RELATIVITY AND LIGHT

Everything you see is in the past. It takes time for light to travel from point A to point B, even if it travels at 186,000 miles per second. This may seem fast, but it's quite slow in a massive universe.

For instance, the average distance between the Sun and Earth is around

THE SUNSET CONJECTURE

92 million miles. So, light from the sun takes around 8.3 minutes to reach us.[41] This means the light you're receiving from the sun right now is not the same light the sun is emitting *right now*. If the sun exploded at this very moment (which it probably won't), then you wouldn't know for about eight minutes. At which point, it wouldn't really matter since you'd be, as basketball star Shaq says, "BBQ Chicken."

Whenever you receive light (all the time), you are making a measurement about the past. This holds true whether you're looking at a star ten light years away (figure 84-a) or if you're looking at someone sitting ten meters away (figure 84-b). In the first case, you are looking at what that star looked like ten years ago. In the second case, you are looking at what someone looked like 0.00000003 seconds (or 3×10^{-8} seconds) in the past. This concept is one of the defining principles of relativity.

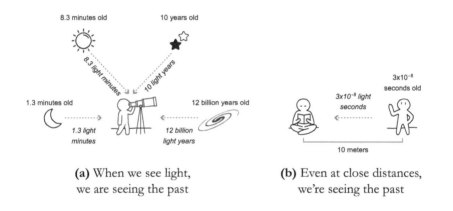

(a) When we see light, we are seeing the past

(b) Even at close distances, we're seeing the past

Figure 84. No matter how far away you're looking, you're always looking in the past.

In the time it takes for the emitted light of an object to reach its destination, that object has moved. For instance, imagine we are looking at the Messier 87 galaxy (figure 85-a). At a time, t_0, the Messier 87 galaxy emits light toward us at 186,000 miles per second. At that speed, it takes around 54 million years to reach us. When the light eventually reaches us, we'll be able to make a measurement at a later time, t_1. There, we'll see Messier 87 as it was in location (x, y, z) in space and t_0 in time. The problem is in the 54 million years it took for light to arrive, the Messier 87 galaxy moved to a new location (x', y', z') in space and t_1 in time. Since it takes time for light to reach us, we won't see what Messier 87 looks like at t_1 for another couple million years. As activist Martin Luther King said, "Everything we see is a shadow cast by that which we do not see."

This concept holds true at any level. For instance, imagine we look at a cell underneath a microscope (figure 85-b). It takes time for light from that cell to reach our eyes. At the point when the light reaches our eye, the cell will have moved to another position. We won't be able to see what that cell looks like in its new position for another split microsecond.

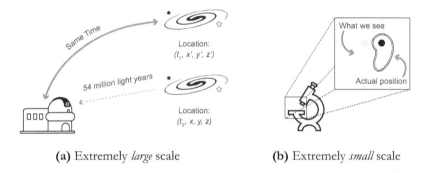

(a) Extremely *large* scale   (b) Extremely *small* scale

Figure 85. Relativity works on both large and small scales: by the time we make a measurement of an object, it has moved to another position.

LIGHT CONES AND WORLDLINES

A *light cone* defines all the possible paths light could take to reach or leave an event, or moment in spacetime (figure 86). For example, let's call this present moment event p. Light cones consist of three different "times" for an event. The bottom part of the cone consists of everything in your past; the top part is everything in your future; and the intersection of the two is the present, or you now. Here, it is important to remember the wise words of Joseph Goldstein, "You experience the past as a thought in the present. You experience the future as a thought in the present." In reality, there is only the present.

THE SUNSET CONJECTURE

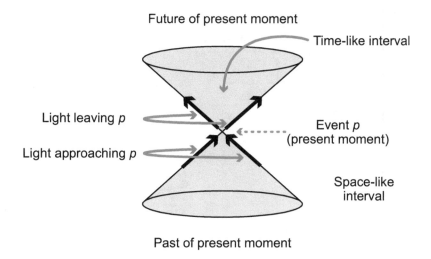

Figure 86. Our present moment, event p, consists of all the light approaching you now and all the light you are emitting now.[42]

Everything that happens within the light cone is considered a *time-like interval*, because it moves slower than the speed of light. In other words, a signal in the past of our present moment can reach us by moving at a speed less than the speed of light. Therefore, any events in that region can affect our present moment. Similarly, a signal emitted in our present moment can affect any event within the time-like future of our present moment.

For instance, when you stand still, you go from event o to event p (figure 87-a). This is an example of a *worldline*, or a path a particle or human takes through spacetime. In this example, you are at the same point in space, only at a different time. What happened there then affects event p now. If instead someone shot a bullet toward you from event q, then it would reach you at event p (figure 87-b). In both instances, an object or signal in the past reached event p by moving slower than the speed of light.

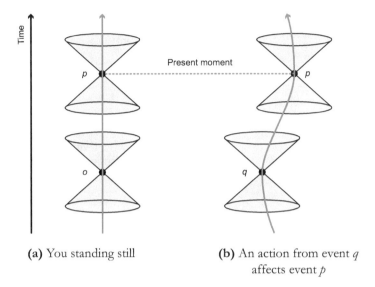

Figure 87. Events in our past light cone can affect our present moment.

Similarly, if you shot a bullet at event p, then it would have an effect on your future world and could affect an event r (figure 88). These are examples of "time-like" intervals: you could reach any event within the light cone in a certain amount of time as long as you move slower than the speed of light.

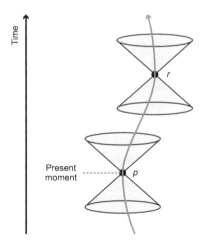

Figure 88. What you do in the present moment affects everything within your future time cone.

On the other hand, everything outside the light cone is considered a *space-like interval*. Space-like intervals are neither in the past nor future of your present moment. To reach an event here, you would need to travel faster than light. For example, let's call the sun exploding, event *s*. This event does not affect your present moment, now, because it takes time for the effects of the explosion to reach you. It is neither in your future nor in your past. But what happens at event s can affect you *later* at event *p'* (figure 89). The sun exploding doesn't reach you now, but it will reach you later.

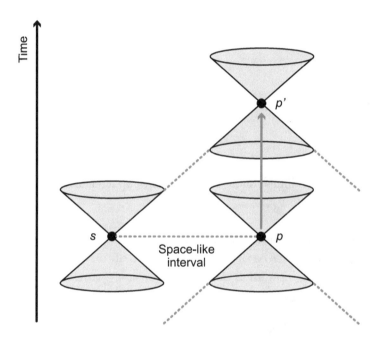

Figure 89. The sun exploding doesn't affect us now, but can affect us later.

To summarize, worldlines can move in time-like intervals, but not space-like intervals (figure 90). In order to travel in a space-like interval, you would need to move faster than the speed of light. This is currently not possible. Sadly, there is no "Beam me up, Scotty." Of course, this would be different if we were made up of *tachyons*—a completely hypothetical particle that can only travel faster than the speed of light.[43] But the only place were tachyons currently exist is in the fertile imagination of *sapiens*.

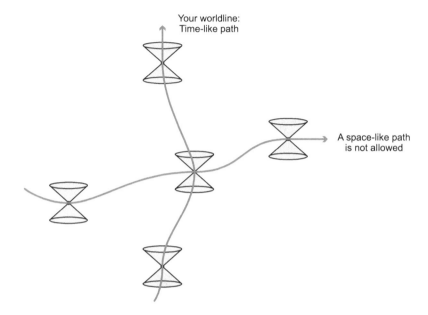

Figure 90. Worldlines follow time-like paths. Space-like intervals are not allowed.

Katie Mack explains light cones much more intuitively in her amazing book, *The End of Everything*. Assume a sheet of two-dimensional space with time moving upward as our third dimension (figure 91). Here, we see a galaxy whose light has traveled 12 billion light years as it was 12 billion years ago; a star whose light has traveled 10 light years as it was 10 light years ago; the sun whose light has traveled eight light minutes as it was eight light minutes ago; etc. The bottom part of our light cone consists of objects close enough to us that the light of the object has had time to reach us. If you want to see what the galaxy whose light took 12 billion light years to get here looks like now, then you would have to travel in a "space-like" manner, and leave the light-cone, to get there.

From here, a natural question arises:

Okay, that's a three-dimensional light cone—two spatial dimensions and a time dimension. What does a four-dimensional light cone—three spatial dimensions and a time dimension—look like?

THE SUNSET CONJECTURE

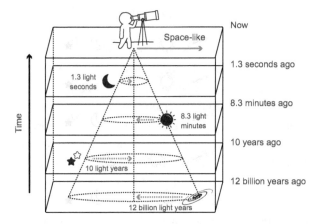

Figure 91. In a two-dimensional space, the bottom of the light cone is what we see in the past, while the tip of the light cone is the present.[44]

Well, you tell me! The answer is everything you're seeing right now! In other words, your *observable universe*, which is everything you can currently see, consists of the edges of your past light cone moving upward through time (figure 92). You are always looking back in time! Every day, your observable universe grows larger and larger as you learn more and more about the past. Furthermore, at the center of your observable universe lies your idea space.

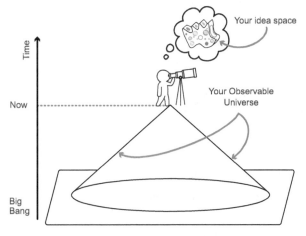

Figure 92. Your observable universe is the edge of your past light cone moving through time.

EDGES OF OUR UNIVERSE

As we saw in the fractals chapter, it is hard to pinpoint exactly where you start and where you end. In reality, your Non-Self is the amalgamation of all your unique layers of your own observable universe—everything from the science of objects and the science of the first person. For these reasons, it is important to understand the structure of your observable universe and what all your layers are, especially the ones on your outer edge.

Our universe has two edges: (1) the Big Bang, the beginning of our own observable universe, and (2) the Hubble Radius, which is the limit of our universe, now. Understanding these limits will help demonstrate how unique your perspective of the universe actually is. Let's start with the first, then move to the second.

Your observable universe is a giant sphere centered on you. Every direction you look in is considered the past (figure 93). It took light from Saturn 90 minutes to reach you; 24,000 light years for light from the center of the Milky Way to reach you; and 13.8 billion years for light from the residue of the Big Bang to reach you. Even the book you're reading right now is considered the past!

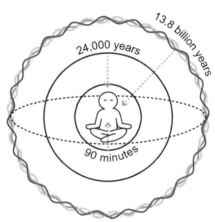

Figure 93. Your observable universe is a giant sphere of past events centered on you (not drawn to scale).

Although the universe as a whole has no clear center, everyone is at the center of their own observable universe. So, everyone's perception of the observable universe—from a butterfly ten meters away to the Big Bang

THE SUNSET CONJECTURE

singularity at the edge of the observable universe—is inherently unique.

Simply put, since everything you see is in the past, the edge of your observable universe represents its beginning, the Big Bang, or the beginning of time; and, the beginning of *your* universe is slightly different than the beginning of *my* universe. In other words, similarly to how a sun simultaneously sets differently on two people, so too does the Big Bang set differently on two people. Picture two giant spheres that overlap for the most part, except for a tiny region on the outside (figure 94). That tiny region on the edge is the Big Bang. Hence the name Singularity Sunset: everyone experiences their own beginning of their own observable universe; everyone experiences their own beginning of time.

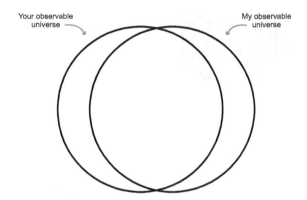

Figure 94. Two observable universes have a large area of overlap, but the outer edge, the Big Bang, is unique to the individual.

To be clear, the Big Bang did not just happen, but it is still happening in our own observable universe! Of course, the Big Bang itself is an event of the past; but, since everything we see is in the past, the effects of the Big Bang are still alive and well, even to this day! As each day passes, your observable universe grows more and more, as new light comes pouring in to reveal more about the past and the beginning of time.

Another name for the edge of our observable universe, or the Big Bang, which is responsible for the constant creation of our own universe, is what musical guru Rick Rubin unknowingly calls the "Source."[45] Overall, both of our observable universes are still being created and *your* beginning is slightly different than *my* beginning. We each have our own "Source."

In principle, you should be able to see all the way back to the edge of your observable universe, or to the "Source."[46] Sadly, that's not the case. There's a giant veil, called the *Cosmic Microwave Background* (CMB), which prevents us from peering further back in time.

The CMB was our universe around 380,000 years after the Big Bang. Before that time, the universe was *very* hot and dense. Light wasn't able to move around freely and kept colliding with all the particles around it. Thus, the universe was opaque. Suddenly, 380,000 years after the Big Bang, the universe expanded and cooled enough to turn transparent. In other words, light was able to move freely through space in a process called *recombination* (figure 95).

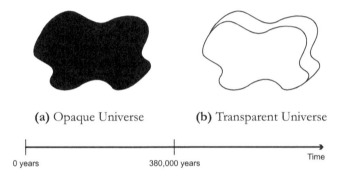

Figure 95. The Cosmic Microwave Background (CMB) is when the universe went from opaque to transparent.

The actual CMB looks like figure 96. Basically, in whichever direction you look you'll see light from the CMB reaching you. During recombination, or 380,000 years after the Big Bang, the average temperature of the universe was around 3,000 Kelvin (K).* Flash forward to today and the temperature has cooled to around 2.7 K with small fluctuations of around 0.002 K, represented by the varying colors in the figure below.[47] Essentially, 2.7 K is the average temperature of outer space.

---

* To get Kelvin, add 273 degrees to the Celsius value (e.g., room temp = 73 F = 23 C = 296 K). 2.7 K is very cold.

THE SUNSET CONJECTURE

Figure 96. What the Cosmic Microwave Background (CMB) looks like. This temperature profile shows a pretty uniform temperature distribution of around 2.7 K.

An interesting feature of the universe is it is *continuously expanding*. In other words, it is impermanent, or in a constant state of flux. This means light emitted from a source four billion light years away actually ends up taking more than four billion light years to reach us! An example will help illustrate this concept:

> Suppose you were alive 380,000 years after the creation of the universe. At this time, every point in the universe has roughly the same temperature and looks like the CMB, or is opaque. The light we observe from the CMB today was emitted from a point that was once 42 million light years away (figure 97-a).
>
> As the universe continued to expand, it gradually cooled down, eventually becoming transparent. This allowed the point 42 million light years away to emit light toward you at a speed of 186,000 miles per second. However, due to expansion of space, light emitted from that point took *more* than 42 million light years to reach you! This is because every point in the universe is constantly expanding away from every other point.[48] In fact, it took approximately 13.8 billion years for that light to reach you, and the same point in space that was once 42 million light years away is now about 46 *billion* light years away (figure 97-b)! We cannot see what that new point looks like now, because its light has not reached us yet.

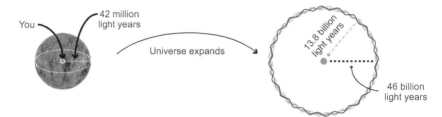

(a) *380,000 years after Big Bang:* Every point in the universe looks like the CMB. Light we see from the CMB today (i.e., 13.8 billion years after the Big Bang) was once 42 million light years away

(b) *13.8 billion years after Big Bang (today):* Light from a point once 42 million light years away has finally reached us after travelling for 13.8 billion years. That point has expanded from a distance of 42 million light years away to 46 billion light years away

Figure 97. Light from the CMB is light that was once 42 million light years away but traveled 13.8 billion light years to reach us. Today, that same point is 46 billion light years away.

What causes this universal expansion that makes our universe impermanent? The answer comes under many names, such as *dark energy*, *vacuum energy*, or *the cosmological constant*. No one knows what this energy is other than it has two consistent properties: (a) the density of dark energy is constant at every point in space, and (b) the density of dark energy is constant at every point in time.[49] So, if I measured how much dark energy there is in a particular sphere at one point in time and compared it to an equal sphere at another point in time, then there would be the same amount of dark energy present in both spheres.

This is counterintuitive. For instance, picture a balloon filled with air. If you were to increase the volume or size of the balloon over time without adding any outside air, then the density of air inside the balloon would decrease (figure 98-a).* Now, let's try applying this same logic to our universe. Instead of seeing a decrease in dark energy density as our universe expands, we see the density remains constant (figure 98-b). In

---

* Density is mass per volume (density = mass/volume). If the mass is constant and the volume goes up, then the density must go down.

other words, as we move forward through time, our universe gets made up of more and more dark energy.

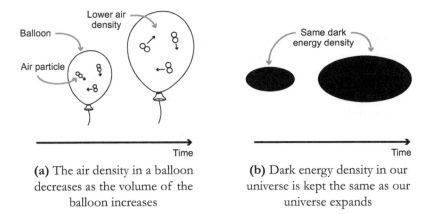

(a) The air density in a balloon decreases as the volume of the balloon increases

(b) Dark energy density in our universe is kept the same as our universe expands

Figure 98. Dark energy density is constant even as space expands.

This phenomenon is weird, because as the universe expands, all other matter and radiation densities get diluted away, similarly to the balloon (figure 99). In a way, it's as if someone, or something, is continuously pouring dark energy into the mold of our universe. Over time, dark energy makes up more and more of our universe.

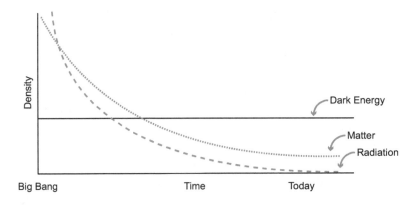

Figure 99. Over time, dark energy makes up more and more of our universe.[50]

The actual value of dark energy density remains an enigma, known as the *cosmological constant problem*. According to observations, the value is around 10^{-15} Joules/cm³.* One centimeter is about the size of your fingernail. So, make a cube with your fingernails and there's 10^{-15} Joules of dark energy there. Do it again at a different point in space and time, and you'll find the same value. Again, this is the observed value. According to quantum field theory, the value should be closer to 10^{105} Joules/cm³. That's a discrepancy of around 10^{120}, or 1 novemtrigintrillion! For comparison, there's about 10^{80} particles in the observable universe and 10^{20} grains of sand on all the beaches on Earth. This discrepancy remains one of the biggest unsolved mysteries in the world of physics.[51]

Universal expansion leads to another interesting property known as *Hubble's Law:* The farther away an object is, the faster it recedes from us. Physicist Kip Thorne explains this concept well in his book *Black Holes & Time Warps*:

> The expansion of the Universe is like the expansion of the surface of a balloon that is being blown up. If a number of ants are standing on the balloon's surface, each ant will see all the other ants move away from him as a result of the balloon's expansion. The farther away another ant is, the faster the first ant will see it move. Similarly, the farther away a distant object is from Earth, the faster we on Earth will see it move as a result of the Universe's expansion. In other words, the object's speed is proportional to its distance.[52]

This would explain why the majority of galaxies, other than our local group, seem to be moving away from us with the farther ones receding faster than the ones closer to us.

This continuous expansion helps define the other edge of our universe, the *Hubble Radius*. As stated above, the farther away something is from us, the faster it recedes from us. Therefore, there will be a point where *the expansion of space itself moves faster than the speed of light*. In other words, an

---

* A Joule is a measure of energy. For reference, the atom bomb at Hiroshima released about 1.5×10^{13} Joules, while a light bulb releases about 60 Joules of energy per second. So, dark energy density is very small!

object that emits a signal beyond the Hubble Radius will forever be unseen. Current estimates place the Hubble Radius at around 14 billion light years.[53]

As an example, you'll be able to see a star moving away from you within the Hubble Radius (figure 100-a); however, as soon as that star passes the magical radius, it will be lost for an eternity (figure 100-b). Light from that star will never reach you since the space between you and the star is expanding faster than the speed of light. Remember the point that was once 42 million light years away and is now 46 billion light years away? Well, since that point is now 46 billion light years away, it is outside your Hubble Radius and you'll never see it again! According to Fermilab, about 20,000 stars leave our Hubble Radius per second.[54]

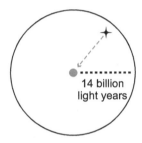 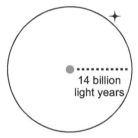

(a) We see light currently emitted from objects within the Hubble radius

(b) We do not see light currently emitted from an object outside the Hubble radius

Figure 100. Any object outside our Hubble Radius moves away from us faster than the speed of light; and, therefore, the light it emits will never be able to reach us.

The two boundaries of our universe are the Big Bang in the observable universe and the Hubble Radius in the present universe. The first deals with objects of the past, while the second looks at where objects are now. Put another way, the Hubble Radius deals with objects not in our light cone yet, while the observable universe deals with objects on the edge of our light cone (figure 101). We are just now seeing objects that were once 42 million years away whose light took 13.8 billion years to reach us. Those objects are now 46 billion light years away.

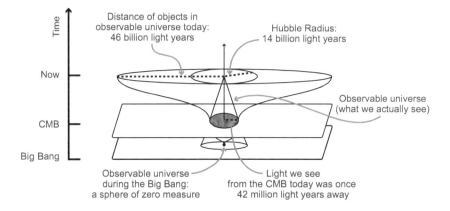

Figure 101. The observable universe deals with our past, while the Hubble Radius defines the edge of our current universe.

Just as the fractals chapter highlighted the difficulty of determining where you start and where you end, the observable universe adds yet another layer of complexity and uniqueness to each individual's Non-Self. By understanding the structure of our observable universe and the ever-expanding nature of the cosmos, we can better appreciate how truly unique our perspectives are. Our individual observable universes, influenced by the cosmic dance of expansion and the mysterious presence of dark energy, shape our understanding of the universe and the interconnectedness of everything within it. By acknowledging our unique vantage points in this vast cosmic landscape, we can gain a greater appreciation of our place in the universe and the intricate beauty of its seemingly infinite layers.

THE SUNSET CONJECTURE

Come back to your beach. Picture the sun's golden rays setting ever so gently right toward you. Although it took eight minutes to reach you, no one else can get the same perspective of that sunset as you (figure 102-a). Now, what if we extrapolate this to the observable universe? No one else receives light from Saturn the same as you. Your own Saturn Sunset. No one receives light from the center of the Milky Way the same as you. Your own Milky Way Sunset. No one receives light from the Cosmic Microwave Background the same as you. Your own CMB Sunset. Finally, no one receives "light" from the Big Bang the same as you. Your own *Singularity Sunset* (figure 102-b).

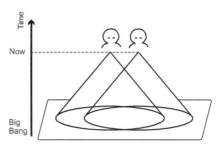

(a) During a sunset, the sun's golden rays appear unique for both people

(b) Everyone experiences their own Singularity Sunset

Figure 102. Everyone experiences their own Singularity Sunset.

I say "light" because actual light from the Big Bang era is trapped behind the veil of the CMB. Although light cannot get past, other particles, like the hypothetical *graviton*—the particle thought to be responsible for gravity—can. In other words, everyone experiences the gravitational effects of the beginning of time, the Big Bang, uniquely.[55]

This is what is meant by your own Singularity Sunset: everyone experiences all the layers of their own observable universe, from its beginning (Big Bang) to the present moment (idea space), differently. Everyone experiences their own "Source." Furthermore, it is impossible to identify where you start and where you end, so your Non-Self is made up of all the distinctive layers of your observable universe. Therefore, since every layer of everyone's observable universe is unique them, everyone experiences their own, unique Non-Self.

All in all, the Sunset Conjecture is two-fold: (1) you are at the center of your own observable universe with your own Singularity Sunset (figure 103-a) and (2) at the center lies your idea space—a topological singularity of zero measure and uncountable depth, hidden from the outside world (figure 103-b).* In the second diagram, time is indicated downward to

---

* Although figure 103-b looks similar to an embedding diagram, covered in the next chapter, it is not the same. It is more so an illustration to show we are falling through time, and at the center of our observable universe lies a topological singularity.

demonstrate that everything you see is in the past. In a way, we're really all just falling through time (figure 103-c).

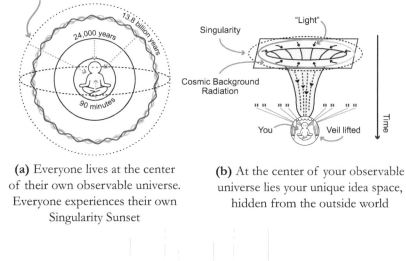

(a) Everyone lives at the center of their own observable universe. Everyone experiences their own Singularity Sunset

(b) At the center of your observable universe lies your unique idea space, hidden from the outside world

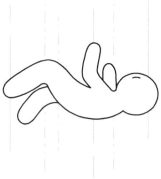

(c) We're all falling through time

Figure 103. The Sunset Conjecture.

As a short aside, there is hidden beauty in the second part of the conjecture. Since your idea space has zero measure, it challenges our traditional understanding of space and time. Space and time describe a framework for things we can measure. For instance, you can measure two feet here or 30 seconds there (figure 104-a). If you were to place an object with zero measure, like a topological singularity, in the way, then you'd still get the same two feet here and 30 seconds there (figure 104-b). An object with zero measure has no effect on the total measurement. Put another away, imagine a car traveling 100 feet in three seconds. If I place an object of zero measure directly in the way of the car, it wouldn't affect the car at all. Therefore, objects of zero measure seem to defy our conventional notions of space and time, as they have no direct effect on neither space nor time. This leads to an interesting question: *Does an object of zero measure sit outside of spacetime?*

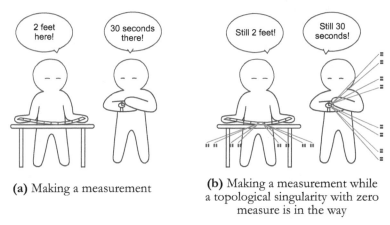

(a) Making a measurement

(b) Making a measurement while a topological singularity with zero measure is in the way

Figure 104. Objects with zero measure sit outside of space and time, because they have no direct effect on neither space nor time.

As Karl Popper writes in regard to Immanuel Kant's writings in the classic *Conjectures and Refutations*:

> Kant concluded that our ideas of space and time are inapplicable to the universe as a whole. We can, of course, apply the ideas of space and time to ordinary physical things and physical events. But space and time themselves are neither things nor events: they cannot even be observed: they are more elusive. They are a kind of framework for things and events . . . part of our mental outfit, our apparatus for grasping this world . . . This is why we get into trouble if we misapply the ideas of space and time by using them in a field which transcends all possible experience.[56]

In other words, space and time are filled with real objects: stars, cells, bodies, etc. But there are objects, like your idea space, for which the framework of space and time seems inadequate, because these objects do not interact with spacetime in a traditional manner. In a way, as we shall see in Chapter 10, space and time are creations of our macro idea spaces. They are a human creation used to describe the world around us. That said, space and time are probably the most useful creations of our macro idea space to date.

To summarize, although our idea space has no direct effect on spacetime, there is a real connection between an idea space and the universe: the body. Parts of our idea space can demonstrate itself through words, body language, or even chemical signals in the brain. However, at the end of the day, no one can perfectly see all your thoughts, emotions, sensations, and perceptions. No one can see your unique perspective of the universe, your Singularity Sunset. Your headless world. To others it's all hidden. It all looks like nothing.

The Sunset Conjecture is the bridge between the science of objects and the science of the first person (figure 105). It is the connection between what can be measured and the infinity hidden beyond the veil of nothingness. In an impermanent universe, where you are constantly tasked to measure facts about the past, lives an immeasurable, impermanent present. A present unique to you. Your Non-Self. Your own Headless Way. What is the present if everything you see is the past?

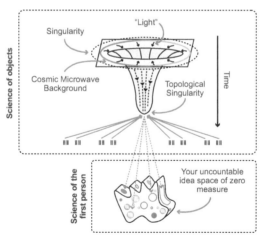

Figure 105. The Sunset Conjecture is the bridge between the science of objects and the science of the first person.

THE SUNSET CONJECTURE

Overall, the Sunset Conjecture lifts the next veil of illusion on our Path of Awakening (figure 106). Prior, everything affected spacetime and occurred *now*. Now, your idea space challenges our preconceptions of space and time and everything you see is in the past. The words of poet David Whyte beautifully capture the essence of this veil: "Our genius is to understand and stand beneath the set of stars present at our birth. And from that place, seek the hidden single star, over the night horizon, we did not know we were following."

After building a framework for the mind, we're now able to see how the mind relates to the universe at large. In the previous chapter, we saw it is impossible to pinpoint exactly where you start and where you end. We are essentially a combination of the science of the first person and the science of objects; our inner and outer worlds. This led us down a path to explore the science of objects in cosmology and the structure of the universe. Here, we saw everyone lives at the center of their own observable universe where everything you see is in the past. And at the center lies your idea space of uncountable depth and zero measure, hidden from the outside world.

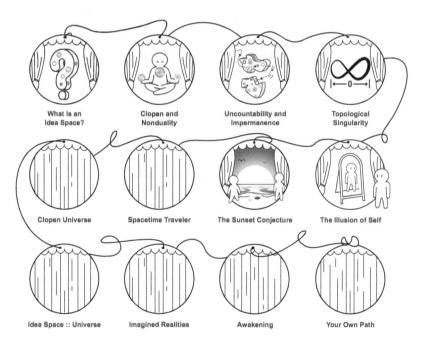

Figure 106. The Sunset Conjecture lifts the next veil of illusion on our Path of Awakening.

As you go through life, it's important to remember that others also experience their own Singularity Sunset, living in the same blissful, headless condition. By assuming good intentions, you can avoid creating fictitious scenarios about what could have been and instead attune to what is here, now. Practice mindfulness both internally, to appreciate your own sunset, and externally, to recognize that others have their unique perspectives as well. In the ever-changing tapestry of life, we are all artists painting our own unique sunsets—masterpieces of perception that remind us of the profound beauty and depth that exists within each and every one of us (figure 107).

Figure 107. A painting the author made of his own sunset.

Chapter 7
MOVEMENT

"Think lightly of yourself and deeply of the world."
- Miyamoto Musashi

Take a deep breath in . . . Let's begin by doing some simple movement. Take your arm and gently wave it back and forth. What does that feel like? What sensations arise? What do you notice? If you're sitting, slowly stand; if you're standing, slowly sit. What do you notice as you transition? Does time slow down? Now focus on the breath. Where does it go in your body? Where do you feel it the most? What do the sensations in your chest and belly feel like?

The connection between your idea space and the universe is your body, and the relationship between mind-body, or *nāmarūpa* (Sanskrit), is clopen. Are we controlled by our body, filled with trillions of cells and bacteria? Or are we controlled by our idea space, filled with uncountable thoughts, emotions, sensations, and perceptions? Both. The situation is clopen (figure 108). Depending on the time of day, you will be more present in one than the other.

Mindfulness of body, especially body postures, provides a tangible source that your awareness can focus on. It grounds you in the awareness of the body instead of being carried away by the mind. Overall, there are four main postures: sitting, walking, standing, and lying down. To practice mindfulness of body, when sitting, you note "sitting," when walking, you note "walking," etc. It does not have to be complicated. As Scott Mescudi, aka musician Kid Cudi, once said, "The ones that make it complicated never get congratulated."

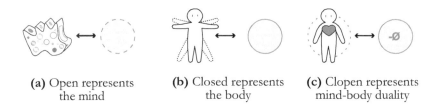

(a) Open represents the mind

(b) Closed represents the body

(c) Clopen represents mind-body duality

Figure 108. Mind-body duality is a clopen experience.

The body provides a perfect object for meditation. It gives us an escape, should we need it, from the restless mind. The beauty is mindfulness of our body, and its movements, runs deep. For instance, you can simply note "sitting." That's the surface level. Can you go deeper? What sensations arise when you sit? Pressure on your feet against the ground? Perhaps your butt on the chair or cushion. Are you completely stationary or are there any micromovements? Can you feel your blood rushing through your body? Perhaps you notice your eyes moving in their sockets as you read this sentence. There is a hidden depth to movement that this chapter aims to uncover.

The next veil of illusion we'll lift on our Path of Awakening is *everyone is a time traveler*—just not in the way you're probably thinking. As we shall see, when you move in a direction, your body contracts in that direction relative to a stationary observer. To make up for that spatial contraction, your time slows down relative to the same stationary observer. For example, if I run really fast (close to the speed of light) in one direction, then, from your perspective, it will look as if my body is contracting in the direction I'm running in, and it will appear as if my watch runs slower than your watch. This effect is perceptible to observers only at very high speeds, but it happens at a miniscule level at normal human speeds, too.

Time is personal. From this simple fact we see two things: (1) space and time are one and the same, hence *spacetime*, and (2) spacetime is malleable. Simply by moving, breathing, loving, laughing, crying, smiling, and having a beating heart we change the world around us. Our direct impact on the world around us via our body may seem trivial, but the goal of this chapter is to show you this fact from a much deeper level.

To do so, we'll see exactly how your body moves through spacetime. We'll start by detailing the geometry of spacetime on a macro level, which is commonly known as *general relativity*. This will show you how gravity, or the illusion of gravity, makes time personal to you. In other words, everyone experiences their own *proper time*, which is the time elapsed by an observer moving in a straight path between two events, or points in spacetime. Afterward, we'll look at how your body moves through spacetime on a micro level through *special relativity*.

Understanding how your body moves through spacetime is important, because your body plays quite a large role in relation to your idea space. Having an understanding of how your body moves in the ether provides an objective view of reality and helps build a truer understanding of reality, which is key to awakening. Furthermore, the geometry introduced in this chapter will play a vital role in Chapter 9 when we explore the structure of your idea space.

GEOMETRY OF SPACETIME

General relativity describes the warping of spacetime to create the illusion of gravity. Intuitively, everything with mass, or energy, bends space around it.* Picture a trampoline. If you place a bowling ball at the center of the trampoline, then the trampoline will dip downward—the curvature and geometry of the trampoline changes. If you then throw marbles around the trampoline, they will tend to fall, or *gravitate*, toward the bowling ball. In this example, spacetime is the trampoline, and matter and energy are the bowling ball and marbles. If you put matter on spacetime, then the geometry of spacetime changes. The larger the matter, the more spacetime curves, and the larger the gravitational effects. In reality, gravity is simply an expression of geometry, as we shall see more clearly in the next section.

Overall, as physicist Steven Weinberg suggests in his book *Gravitation and Cosmology*, "In my view, it is much more useful to regard general relativity [or the curvature of spacetime] above all as a theory of *gravitation* . . ."[57] Or

---

* Mass and energy are one and the same. Think: $E = mc^2$. Energy equals mass times the speed of light squared.

as physicist James Hartle simply states, "Gravity *is* geometry."[58] In other words, similarly to how Isaac Newton's theory of gravity replaced the "law of attraction," so too does Einstein's theory of general relativity replace Newton's theory of gravity.

The fact that our understanding of gravity has evolved over time demonstrates scientific theories are expressions of our macro idea spaces, reflecting our ever-changing perspectives and insights, as we shall see in Chapter 10. Recognizing the fluidity of scientific theories is crucial because it reminds us that theories are not definite truths that should be taken at face value, but rather evolving frameworks for understanding the world around us. In truth, every theory has a limit, which inherently defines the theory itself.

To better understand the properties of spacetime, we must expand our knowledge of Non-Euclidean geometry. Euclidean geometry relates to flat surfaces and basic shapes, like lines, squares, cubes, circles. It's the geometry we're most used to and use every day. Non-Euclidean geometry relates to curved surfaces and fractals.

In the context of geometry, the term "flat" refers to a surface or space without any curvature, where the familiar rules of Euclidean geometry apply. In everyday life, we tend to think of flat surfaces as being two-dimensional, like a sheet of paper or a tabletop. However, in the broader realm of geometry, "flat" can also apply to three-dimensional or even four-dimensional spaces that lack curvature.

It may be challenging to envision this concept since we're used to experiencing gravity and other forces that cause curvature in our everyday lives. However, imagine an idealized scenario where there is no gravity or any other force acting upon a 3D space, such as an infinitely large room in a hypothetical universe without any form of mass or energy. In such a space, the rules of Euclidean geometry would hold true, and the space itself would be considered "flat" in three dimensions.

The conundrum of our universe is two-fold. First, although everything is locally flat, spacetime as a whole has global curvature (more on this soon). Second, perfect lines, perfect squares, perfect cubes, and perfect circles are seldom found in the world. From afar, they sure look straight, but if we zoom in more and more, a fractal world reveals itself with no perfect lines, squares, cubes, and circles in sight (figure 109). Think back to the question: *How long is the coast of Great Britain?* Does true level actually exist? In reality, perfect shapes are idealizations we use to better approximate the world around us. And, at the end of the day, life is but an approximation.

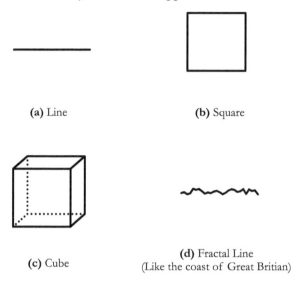

Figure 109. (a-c) Euclidean geometry deals with idealized shapes. **(d)** Non-Euclidean geometry represents a more accurate viewing of our world.

Let's do an experiment to better understand the differences between Euclidean and Non-Euclidean geometry:

> First, draw a line with a pencil. Any line will do. Then, draw a circle around the line, so the line becomes the *diameter* of the circle (figure 110-a). Now, eyeball the length of the circle, or its *circumference*.* How long does the circumference look? What you have there is Euclidean geometry. Now, push the pencil into the middle of the circle so the paper curves (figure 110-b). This is Non-Euclidean geometry. Eyeball the circumference of the circle on your curved paper once more. Does the circumference look shorter, longer, or the same?

* The radius is half the diameter. So, the circumference of a circle is equal to: $2\pi \times$ radius, or $\pi \times$ diameter.

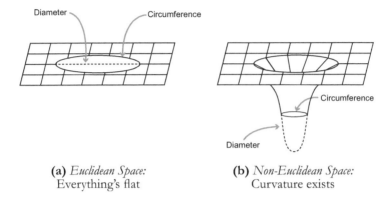

Figure 110. What happens when you **(a)** draw a line and circle, and **(b)** push a pen through that line and circle.

In my experience, the circumference of the circle in the flat, Euclidean space is *larger* than the circumference of the circle in the curved, Non-Euclidean space. Namely, the circumference of the circle on the left follows our beloved equation, pi times the diameter, while the circumference of the circle on the right shrinks to a value less than pi times a diameter! These figures are examples of *embedding diagrams*, and they are important, because they help demonstrate what happens around large stellar objects[59] and how ideas form in our idea space (see Chapter 9).*

Overall, Euclidean geometry is the geometry of flat surfaces (figure 111-a). Here, parallel lines remain parallel, the shortest path between two points is a straight line, etc. The Non-Euclidean geometry from the example above is called *spherical geometry* (figure 111-b). Here, parallel lines eventually converge, and the shortest path between two points is no longer a straight line, but a piece of a great circle. Another example of Non-Euclidean geometry that exists is called *hyperbolic geometry* (figure 111-c). This is often represented by a saddle, or a Pringle's chip. In this geometry, parallel lines diverge from each other, and the shortest path between two points is a hyperbola. Independent of geometry, the shortest distance between any events is the path of least resistance, and it is called a *geodesic*.[60]

---

* In figure 103 of the last chapter, we used a similar drawing to convey the Sunset Conjecture. For clarity, that was not an embedding diagram.

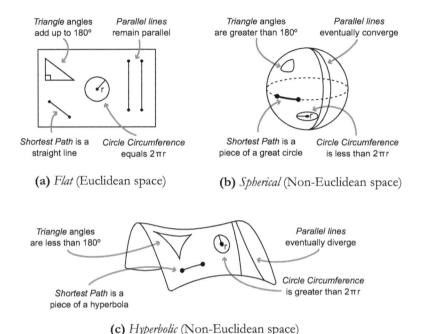

Figure 111. The various types of curvature and associated properties.

To better understand embedding diagrams, we need to make a couple of stops in imagination land. First stop: *Flatland*. The concept, first introduced by Edwin Abbott, goes as follows: Imagine creatures, creatively called flatlanders, living in a universe with only two spatial dimensions, or 2D (figure 112).[61] Clearly, there is some curvature about their universe: a curved bowl in the middle and flatness outside of it. Since they are 2D, can they tell whether there is curvature or not?

Let's say that these flatlanders are you—clearly a smart person! One day, you believe there's more to life than this 2D world and decide to make some geometric measurements. You start by examining a pair of parallel lines, L1 and L2, in the flat portion of the universe. The lines seem to stay parallel no matter how far you go. Nothing fishy there. Next, you decide to pick another pair of parallel lines, L3 and L4, to double-check. You can never be too sure! This time around, something fascinating happens. Lines 3 and 4 eventually meet! How can this be?![62]

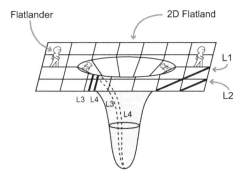

Figure 112. A 2D creature, or flatlander, can test the curvature of his or her world by performing measurements on its geometry.

Of course, this is all with a 2D creature. We live in a three-dimensional world! When we take this example to our planet, Earth, something similar holds true. As you may know, Earth has some curvature to it when looked at from space. However, for us folks living on Earth, wherever we walk, the world *appears* to be flat, just like for the flatlanders! The catch: simply because the world appears flat, does not mean it is flat! For instance, if I walked in a straight line around the equator, then I would make it all the way around Earth and finish exactly where I started. So, even though I walked in what appeared to be a straight line, I actually walked around in a great circle—characteristics of spherical curvature!

The following thought experiment by the illustrious Kip Thorne confirms the spherical curvature of Earth:

> Imagine yourself standing on the North Pole holding two balls. You decide to throw them up in the air—at precisely parallel trajectories—and watch them fall back to Earth (figure 113). Now, the balls have this special property: they are made of a material that falls through the Earth without being slowed at all. Pretend you and your friend, Superman, who's on the opposite side of the Earth, follow the balls' motion as they continue down their path with "X-ray" vision. As the experiment continues, you both notice the balls eventually collide with each other near the center of the

Earth, even though the balls were initially parallel to each other. Thus proving, like the flatlanders did, there must be some curvature.[63]

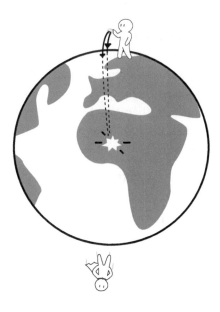

Figure 113. Kip Thorne's thought experiment shows two balls thrown at parallel trajectories will eventually meet near the center of Earth. So, Earth has curvature.

The conundrum is: *How can there be global curvature, yet everything still looks flat?* To answer this question, we turn to Einstein's Equivalence Principle, which states, "In small enough regions of spacetime, the laws of physics reduce to those of special relativity."[64] Special relativity means a flat, or Euclidean, space. We'll touch on the properties of special relativity more shortly; but, for our purposes, this means that, locally, every point in spacetime appears Euclidean, or flat, but spacetime as a whole appears Non-Euclidean, or has curvature. This is a clopen phenomenon: spacetime is flat everywhere, yet curves.

To develop a more intuitive understanding, we turn to another object that is locally flat, yet has global curvature: the *Devil's Staircase*. Imagine walking in a flat, straight line. Then, after ten steps you turn around and notice you've actually climbed upward. In other words, the world appears

Euclidean, but is actually Non-Euclidean on larger scales. This is the Devil's Staircase: wherever you walk, it will look flat, yet you actually changed elevation. In a way, it's as if you were walking up a staircase without even noticing you were walking upward! It must be the work of the Devil! Simply put, the curvature of spacetime is perceptible from a large-enough perspective and not perceptible from a small perspective.

Constructing the Devil's Staircase follows a very similar logic to the Cantor Set. Essentially, we have a function that follows the below schema. Here, the "∈" symbol means "belongs to." For example, the first line reads: if x is between 3/9 and 6/9, then y is 1/2. The second line reads: if x is between 1/9 and 2/9, then y is 1/4. You can repeat this pattern indefinitely and get something like figure 114-a and figure 114-b.*

$$y = \begin{cases} 1/2 \text{ if } x \in [3/9, 6/9] \\ 1/4 \text{ if } x \in [1/9, 2/9] \\ 3/4 \text{ if } x \in [7/9, 8/9] \\ \ldots \end{cases}$$

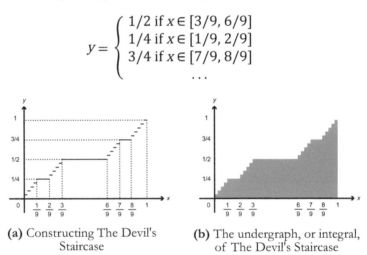

(a) Constructing The Devil's Staircase

(b) The undergraph, or integral, of The Devil's Staircase

Figure 114. The Devil's Staircase is an object that is locally flat, yet has global curvature.

As you can tell, every part of the graph is flat. Yet, we clearly see the graph goes from zero to one. The fabric of spacetime follows a similar trend where every point is flat; however, the overall fabric of spacetime has complex curvature (figure 115-a). For instance, I walk in a straight line around the Earth, yet end up where I started.

---

* Fun fact: the Devil's Staircase is continuous and has a derivative of zero almost everywhere. This is odd, because a derivative of zero is usually indicative of a flat surface, yet this beautiful monstrocity goes from 0 to 1!

MOVEMENT

Practically, this means any point in spacetime can be represented by a reference frame. Look straight. That's x. Look right. That's y. Look up. That's z. Imagine infinite, straight lines extending in each direction. That's a reference frame. Your reference frame is valid anywhere and everywhere in this universe (figure 115-b).

All in all, wherever you are in the universe, everything will look flat. But beware, flat Earthers! Behind the flatness lies a hidden geometry—you are but a Flatlander living in a 3D world.

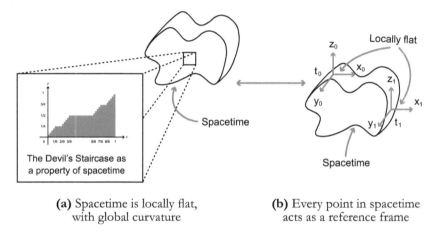

(a) Spacetime is locally flat, with global curvature.

(b) Every point in spacetime acts as a reference frame.

Figure 115. Spacetime is flat everywhere, yet has global curvature, like The Devil's Staircase.

Understanding the geometry of spacetime is important, because it allows us to build a better understanding of the world we inhabit. Furthermore, many of the properties of spacetime translate well into how we build the large-scale structure of your idea space, as we shall see later more concretely in Chapter 9.

GRAVITY AS A PSEUDO FORCE

A natural question arises next: *What does the curvature of spacetime do?* Many things. Most notably, it creates the illusion of gravity. It also completely changes our perception of how we measure distances in space and time. Let's tackle gravity here and our perception of spacetime in the next section.

To better understand how gravity is a *pseudo force*, let's do a thought experiment. As we saw, you are constantly accelerating toward Earth at $g = 9.81$ m/s². This makes sense, otherwise skydiving wouldn't be as much fun. If you were on another planet, then you would have a different acceleration as you jumped out of a plane. For instance, Jupiter has a gravitational constant of 24.79 m/s². If you were able to jump out of a plane on Jupiter, then you would accelerate downward two and a half times faster than on Earth!

Now, picture yourself in a rocket ship in outer space without any windows. The rocket ship is continuously speeding upward with an acceleration $g = 9.81$ m/s². Since the rocket is accelerating upward at g, you are being pulled downward at g. Think of an elevator. When it initially goes up, you get pulled down; when it initially goes down, you feel like you're getting pulled up. Or, as Newton's Third Law says: for every force, or action, there is an equal and opposite reaction.

Imagine you had lived your entire life on that rocket, and the ship's computer told you this rocket was "Earth." Would you be able to tell if it was lying (figure 116)? In other words, can you tell the difference between being on Earth and a rocket accelerating upwards at the same gravitational constant as Earth? Then, what if the rocket accelerates to 24.79 m/s² and the computer now tells you you're on Jupiter? Are you able to tell you are on Jupiter? There are no windows, so I guess you'll have to take the computer's word for it and hope it's not evil, like HAL from *2001: A Space Odyssey* (1968).

Overall, the resounding answer is no: you cannot tell the difference between the physics of a gravitational frame versus an accelerating frame. Thus, gravity is a *pseudo force*. In other words, what can be considered "gravity" is really you continuously accelerating toward an object, like Earth.

The reason we feel gravity even when we are not moving relative to the Earth is because the Earth's surface exerts a resistant force on us. When we stand on Earth, we are experiencing a gravitational force that pulls us toward the center of the planet. However, the Earth's surface pushes back against us with an equal and opposite force. This force is what prevents us from moving toward the center of the Earth and allows us to remain in place. If we were in space near a massive object, experiencing a gravitational force similar to Earth's gravity, we would then accelerate, or gravitate, toward that object due to the lack of any resistant force.

MOVEMENT

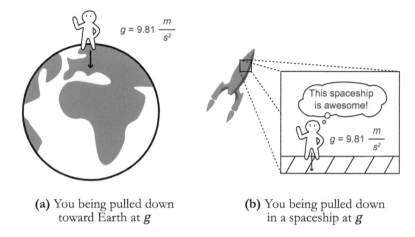

(a) You being pulled down toward Earth at g

(b) You being pulled down in a spaceship at g

Figure 116. Can you tell the difference from being pulled down toward Earth at g versus being on a spaceship accelerating at g? No.

Einstein was the one who had the idea that gravity and acceleration were two sides of the same coin and concluded two principles: The Weak Equivalence Principle (WEP) and The Einstein Equivalence Principle (EEP). The WEP states: The motion of freely falling particles is the same in a gravitational field and a uniformly accelerated frame, in small-enough regions of spacetime. This is the spaceship example. We saw part of the EEP earlier and will explore it more soon. It is a bit stronger than the WEP, which states: In small-enough regions of spacetime, the laws of physics reduce to those of special relativity; it is impossible to detect the existence of a gravitational field by means of local experiments.[65] In other words, spacetime is locally flat and you can't tell the difference between a gravitational frame and accelerating frame.

A last example, this time by the great physicist Richard Feynman, will help seal the deal as to why gravity is a pseudo force:

> Pseudo forces can be illustrated by an interesting experiment in which we push a jar of water along a table, with acceleration. Gravity, of course, acts downward on the water, but because of the horizontal acceleration there is also a pseudo force acting horizontally and in a direction opposite to the acceleration. The resultant of gravity and pseudo force makes an angle with

the vertical, and during the acceleration the surface of the water will be perpendicular to the resultant force, i.e., inclined at an angle with the table, with the water standing higher in the rearward side of the jar. When the push on the jar stops and the jar decelerates because of friction, the pseudo force is reversed, and the water stands higher in the forward side of the jar (figure 117).[66]

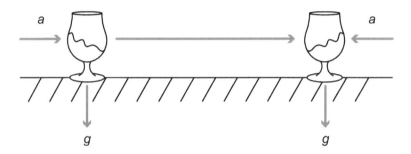

Figure 117. Another example that shows that gravity is a pseudo force.

If you were a tiny sea creature living on the surface of the water with no knowledge of the outside world, would you be able to tell whether the force on the water was due to acceleration or gravity? No.

Einstein's theory of general relativity posits that gravity is not an independent force, but rather a consequence of the curvature of spacetime caused by mass and energy. Objects moving in curved spacetime follow the path of least resistance, or a geodesic, which appears to us as the familiar force of gravity. In this context, acceleration due to gravity is indistinguishable from other forms of acceleration because both involve motion along a geodesic in spacetime. Thus, gravity is not a separate force but rather the manifestation of spacetime curvature.

Simply put, gravity is an illusion caused by the curvature of spacetime. Think of the marbles falling toward the bowling ball on the trampoline. The ball's mass curves the trampoline, so the marbles "gravitate" toward it. In parallel, a star's mass curves spacetime so planets "gravitate" toward the star. This works for any object with mass, or energy.

CURVATURE AND TIME

Depending on where you are in the curvature of spacetime, you will experience a different acceleration, which we commonly interpret as the "gravitational force." This acceleration dictates our perception of how we measure time.

For instance, picture two people around a black hole. A black hole is a region in spacetime where the curvature becomes so extreme that not even light can escape. Essentially, a black hole is a rip in the fabric of spacetime. Imagine pushing down on a piece of paper with a retracted ballpoint pen, like we did before, and then clicking the pen so it breaks through the paper. That's similar to a black hole: it is an object so dense it breaks through spacetime.

The person closer to this massive object will experience a stronger curvature of spacetime and thus a higher acceleration than the person farther away. Remember, the faster the acceleration, the slower time appears to move. Consequently, the person closer to the black hole will perceive time more slowly than the person standing away from the black hole. The clock farther from the black hole appears to run more quickly.

Take Joe and Misty for example (figure 118-a). Here, Joe joyously jumped into a black hole and experienced time more slowly than Misty due to the stronger curvature of spacetime. From Misty's perspective, Joe's clock appears to run more slowly. Where Joe would measure an event taking one second, Misty might perceive that same event to take ten seconds for Joe. They each have their own proper time.

You don't have to be in a black hole to experience this difference. For example, at the time of this writing, I am sitting on the second floor of my house. For me, I perceive time on the second floor to move faster than time on the first floor. This is because the curvature of spacetime weakens as I move farther away from the surface of Earth (figure 118-b).

A last example: As you're reading this, your head is above your feet (I presume). Your head is experiencing time faster than your feet, since the curvature of spacetime is weaker at your head than it is at your feet (figure 118-c). The bubbles in figure 118 represent the perceived elapsed time, or proper time, for each observer. They are for illustrative purposes only.

Sure, the last two examples are "negligible" for calculations, but life is not a calculation. Even though these time differences may seem insignificant in day-to-day life, they highlight a fascinating concept that we are all constantly traveling through space and time at our own unique pace, with the curvature of spacetime shaping our individual journey.

(a) Two people experience different proper times depending on their distance from the black hole

(b) You experience different proper time on the second floor compared to someone on the first floor

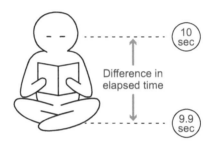

(c) As you read this book, your head experiences different proper time compared to you feet

Figure 118. How fast time moves depends on your gravitational field, or accelerated frame. Please note the numbers are for illustrative purposes only.

SPECIAL RELATIVITY

The above sections demonstrate how general relativity, or the warping of spacetime, affects space and time on a macro scale. However, as Einstein said in his equivalence principle: locally, everything acts by the rules of special relativity. In other words, everything is locally flat, like the Devil's Staircase. Special relativity provides a more practical, day-to-day understanding of how our bodies move through spacetime and offers a clearer perspective of our surroundings and the rules governing our experience. To see it in action, let's explore what happens when one person simply moves relative to a stationary observer.

At a high level, any two people making "the same" measurement will have different results depending on their reference frame.[*] For instance, after two failed attempts with Deep Rule, Joe and Misty decide to go old school and try to measure space with a good old-fashioned *ruler*. The ruler is exactly one meter long.

Joe and Misty heard that funky things can happen when one person moves through spacetime. So, to see what the hype is all about, they decided to test it out for themselves. The experiment: both Joe and Misty will measure a length of one meter in front of them. The catch: Joe will stand still, while Misty will run to the right at a certain speed, or velocity (v_{Misty}).

The two best friends get their rulers ready. Misty starts running toward the right. After a second, they both place their ruler down to measure the length in front of them. After the measurement is made, they regroup to look at the results.

Incredible! From Joe's viewpoint, he made a measurement of one meter, while Misty made a measurement slightly less than one meter (figure 119). In other words, according to Joe, Misty used a foreshortened ruler! From the viewpoint of a stationary observer (Joe), when someone moves in a particular direction (Misty), they spatially contract in that direction of motion. This leads us to a stunning conclusion: space is malleable.

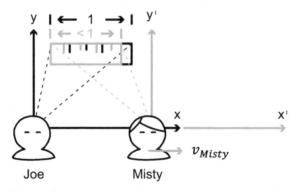

Figure 119. Depending on your reference frame, two people will have different results for "the same" measurement.

---

[*] For a full complementary analysis, then please see the supplemental material at www.TheIdeaSpace.io.

Now what about time? Another example, known as the *Twin's Paradox*, will help illuminate. Imagine two twins born at the *exact the same time*. One day, one twin decides to go into outer space and orbit the Earth at very high speeds. The twin on the ground decides to become a monastic monk and *not move at all*, or stay at rest. Upon the spaceman's return to Earth, the pair decide to get biologically tested to see if they are still the same age. After receiving their results, Dr. Mantis Toboggan explains that the twin who went to space is now *biologically younger than the twin who stayed at rest* (figure 120).

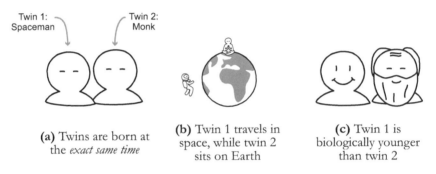

(a) Twins are born at the *exact same time*

(b) Twin 1 travels in space, while twin 2 sits on Earth

(c) Twin 1 is biologically younger than twin 2

Figure 120. When one person moves relative to a stationary observer, their time slows down.

Dr. Toboggan explains, "The twin that went to space was accelerating very fast around the Earth. Since he was moving at a non-constant velocity and had to turn around, his body contracted in the direction he was moving in, relative to us. To make up for that contraction, the spaceman's proper time moved more slowly relative to ours; and, therefore, he didn't age as fast. Basically, *when you move, time slows down for you relative to someone at rest*."[67]

Here, we see why movement is so important: whenever you move, you age more slowly relative to the stationary world around you. Remember this the next time you go on a run or walk: in a way, you move to slow down time—you are a time traveler! As Christopher McDougall unknowingly writes in *Born to Run*, "You don't stop running because you get old; you get old because you stop running." For example, if you spent your entire life on an airplane, then you would have aged merely 0.00005 seconds or so less than your twin.[68] This change in time may be "negligible" for calculations; but, once again, life is not a calculation.

Now, think of every breath you take. As you breathe, your chest inflates and deflates. As it moves, your chest's time slows down relative to your body. No two breaths are the same! It is true that many of these differences are minute, but that's only from our reference frame. Imagine how micro movements affect cells, bacteria, and molecules. For instance, the average lifespan of a bacteria is around 12 hours or so.[69] To them, every breath is a life event. Go one layer deeper and the chemical changes that occur between every breath are too vast to even fathom. One layer more and the changes at the particle level are even more drastic. Time is personal not only to us, but also to all our cells, bacteria, molecules, and particles. *Time is micropersonal.**

So, the next time you exercise, go on a walk, or simply move, see if you can appreciate this small change in temporal perception instead of thinking of all things associated with "I," like "my problems," "I'm tired," etc. As Miyamoto Musashi said, "Think lightly of yourself and deeply of the world."

FAILURE OF SIMULTANEITY AT A DISTANCE

Another interesting phenomena that demonstrates the personal nature of time is called *failure of simultaneity at a distance*. Here, one person sees two clocks to be in synch, while an outside observer sees two clocks to be out of synch. The following thought experiment will illuminate.

Let's go back to our person moving in a spaceship without any windows, and let's call her Lois. Her goal is to place two clocks on either end of the spaceship to see if they are synchronized. The ship is moving at a constant speed, v_{ship}. How will she synchronize the clocks? Lois is smart and decides to place herself in the middle of the spaceship, or at the halfway mark. From this point, she will send out a signal that will go both ways, at the same speed, and arrive at both clocks at the same time, successfully synchronizing the clocks (figure 121-a).

Now, does the same hold true for an outside observer? Say we place an observer with X-ray vision, Clark, outside the spaceship. Does he also perceive the clocks to be in synch? Lois believes they are because she doesn't

---

* A muon particle traveling at 9/10th the speed of light lives for two microseconds from its reference frame. If you were to look at it from a stationary reference frame, then it would live for twice as long! See supplemental material on site.

know the ship is moving. Since there are no windows and the ship is moving at constant speed, she feels as if the ship is standing still. On the other hand, Clark sees the ship moving forward. He deduces the clock in the front end is running away from the signal Lois emits—the signal has to travel *more* than halfway to get to the clock. Alternatively, the rear clock is moving toward the signal, and the signal has to travel *less* than halfway to get to the clock. Thus, from Clark's perspective, Lois's signal reaches the rear clock first, then the front clock (figure 121-b)!

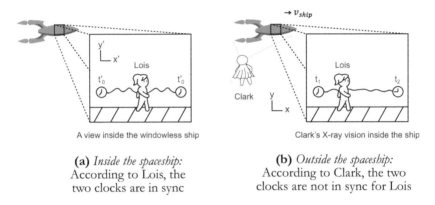

(a) *Inside the spaceship:* According to Lois, the two clocks are in sync

(b) *Outside the spaceship:* According to Clark, the two clocks are not in sync for Lois

Figure 121. Depending on your reference frame, two clocks will be (a) in synch, or (b) out of synch.

Clearly, there is a failure of simultaneity at a distance. Lois sees the clocks in synch. Clark sees the clocks out of synch. Both are true. The situation is clopen. Time is personal.

This thought experiment highlights how different observers, depending on their relative motion, can perceive the same events happening at different times. Just as Clark and Lois disagree on the synchronization of the clocks, the perception of time is personal and dependent on an observer's unique frame of reference.

WORLDLINES REVISITED

Our exploration of spacetime is relevant to your idea space because the more we understand what the world is, the more we uncover what the world is not. This process of discovery defines our awakening. By understanding phenomena like spacetime travel, for example, we lift a veil that once clouded our comprehension of the world.

With this deeper appreciation of space and time, we can revisit worldlines in more detail. Remember, a worldline is a path a particle or human takes through spacetime. What would the worldlines of two people having a face-to-face conversation look like? At first glance, it would look like figure 122. The friends *appear* to be at the same point in time and different points in space. However, as stated earlier, we're all time travelers, just not in the way you're thinking.

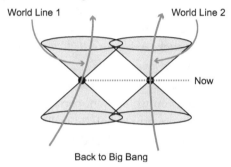

Figure 122. A *false* representation of a face-to-face conversation.

Assuming everyone lives at the same point in time is only an approximation. It is a simplification we use as humans to ensure cooperation on a massive level. Can you imagine how ridiculous it would be if everyone's clocks were different? In reality, spacetime is malleable. Simply moving changes the perception of time for a spacetime traveler (i.e., you). Thus, two people having a face-to-face conversation are not only at a different point in space, but also at a different point in time (i.e., t_1 and t_2), each with their own Singularity Sunset (figure 123).

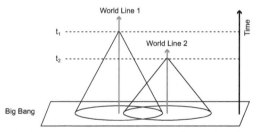

Figure 123. A more realistic representation of a face-to-face conversation.

NAVIGATING DECISIONS AND UNCERTAINTY

As you continue to build your own worldline, you will be faced with decisions you need to make, or *choice points*. In these instances, it's important to take deep breaths and remember the powerful koan by Chao Chou:

The Ultimate Path is without difficulty;
just avoid picking and choosing.

What path feels most natural? What path leads to the least amount of clinging or suffering? If the answer doesn't arise naturally, as it seldom does, use the *Three Me* trick. Ask yourself: *What would past me do, what would present me do, and what would future me do?*

Of course, this depends on what you want for future you! A great tip for deciding what future you would want comes from Naval Ravikant, "Play long-term games with long-term people." Finally, give a completely arbitrary weighting to each of the three "you," and the path will become clear (figure 124).

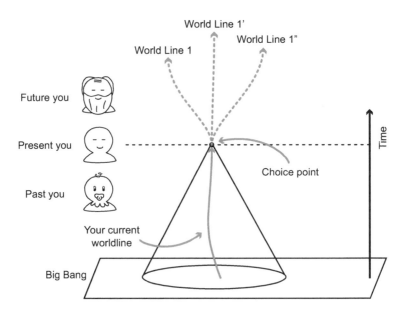

Figure 124. A useful decision-making process for any spacetime explorer, or adventurer: *What would the three me do?*

Sometimes, as you travel in this crazy, wacky world, you'll come across new veils of illusion, or new Unknown Unknowns will enter your idea space (figure 125). These veils characterize themselves in many shapes and forms, like ideas, people, or koans, and act as topological singularities. At first, they look like nothing, ∅, but as soon as they're uncovered, their infinite vastness reveals itself and completely changes your perspective on life.

Be open to lifting the veils. Be forgiving to yourself and others. As Gandhi said, "Freedom is not worth having if it doesn't include the freedom to make mistakes."

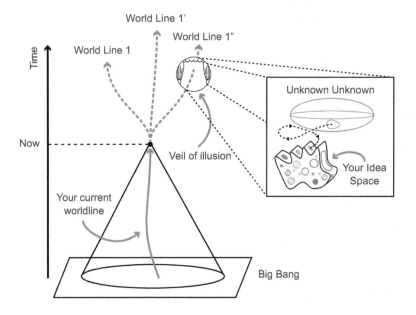

Figure 125. New veils of illusion will uncover themselves as you proceed in life.

As you go through life, it's important to recognize each person navigates their own world line, shaped by their unique idea space. Life is a singular journey, independent and incomparable to any other. There is profound liberation in knowing that each path is its own, gaining wisdom by not falling into the trap of comparisons. Honor your path, embrace its uniqueness, and celebrate its individuality.

TIME IS PERSONAL

So far, we've established a solid framework for describing consciousness in your idea space. We then delved into its properties, exploring the relationship between your idea space and the world around you.

In this chapter, we examined how the body, which intimately connects to your idea space, functions within the ether. We laid the groundwork for understanding the geometry of spacetime, revealing that gravity is simply an illusion caused by the curvature of spacetime. This understanding of Non-Euclidean spacetime geometry sets the stage for Chapter 9, where we'll explore a fascinating parallel: *Could the formation of ideas within your idea space operate in a similar way?*

After the discussion on geometry, we discovered time is not only personal but also *micropersonal*, lifting another veil of illusion on the Path of Awakening (figure 126). Previously, it was believed that everyone lived in the same time. Now, we know that each person experiences their own personal time.

Understanding that each person experiences their own personal time underscores the importance of living from moment to moment. Each moment is unique and unrepeatable, a fleeting snapshot in the ever-changing landscape of our personal time. By fully immersing ourselves in the present moment, we can truly engage with our experiences and navigate our idea space more effectively. This moment-to-moment awareness is not just a mindfulness practice—it's a profound realization of the impermanent and personal nature of our existence.

In the wondrous dance between mind and body, we are both the creators and experiencers of our own reality, shaping and being shaped by the world around us in a harmonious interplay that unites the deepest mysteries of our existence with the tangible, ever-changing canvas of spacetime. As we journey through this intricate landscape, it is important to embrace the beauty of our interconnectedness and the profound implications it holds for understanding ourselves and the universe. By lifting the veils of illusion, we come closer to unveiling the true nature of our existence, where the seemingly separate realms of mind and body converge into a symphony of cosmic proportions.

MOVEMENT

Figure 126. Understanding that your time is personal lifts the next veil of illusion on our Path of Awakening.

Chapter 8

THE CLOPEN NATURE OF REALITY

"There are things known and there are things unknown,
and in between, are the doors of perception."
- Aldous Huxley

Take a few seconds to breathe. Can you solely focus on the first nen, or first thought impulse? What phenomena is present? Is it a sensation? Is it a sound? Is it a sight? Forego the desire to go to the self-observing, second nen that reflects on the experience. *This is my foot. This is a bird. This is a book.* Simply sense. Simply hear. Simply see. Forego the desire to go to the self-conscious, third nen when you become aware, or mindful, of the reflection. *This is mindfulness of sensations. This is mindfulness of sounds. This is mindfulness of sights.* Simply sense. Simply hear. Simply see. Instead of reaching out toward the experience, let it come to you. The world is coming toward you; you are not going toward it.[70] Focus solely on phenomena. Sit in this One-Nen state of mind for a couple of breaths.

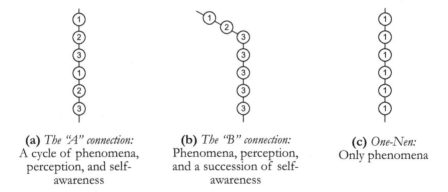

(a) *The "A" connection:* A cycle of phenomena, perception, and self-awareness

(b) *The "B" connection:* Phenomena, perception, and a succession of self-awareness

(c) *One-Nen:* Only phenomena

Figure 127. Different types of nen, or thought impulse, actions.

Clopen represents the coexistence of seemingly opposite ideas, such as open and closed. To experience clopen, recognize the interconnectedness of everything and nothing, as the two are complementary aspects of the same reality (figure 128). For instance, explore the concept of emptiness in your idea space by listening to the silence in between sounds. When you search for the absence of sounds, you become more aware of all the sounds around you. In this moment, you will discover the harmonious interplay between the absence of sound and the presence of all sounds. As an Ancient said, "Silence isn't empty. It's full of answers."

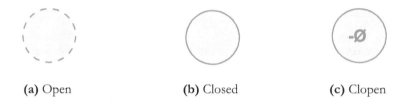

Figure 128. Representations of open, closed, and clopen.

Clopen is a crucial element in awakening, as it provides the necessary structure to understand nonduality. When we categorize objects like thoughts, sensations, emotions, or perceptions, we often create a duality: good vs. bad, hard vs. soft, pleasant vs. unpleasant, this vs. that. These groupings are completely arbitrary and unique to the individual. How you classify one thing is completely different than how someone else does. For instance, think of the famous dress. Two people see different colors in the same dress, illustrating the subjective nature of perception.

The conundrum lies in the fact that grouping is the most foundational human experience. We constantly classify objects—for instance, "good," "bad," "pretty," "ugly," and so on—as grouping supersedes the need to count or even make measurements. If you are not mindful of these groupings, then you can build attachments to them. These attachments then condition the mind to view the world a certain way. As Ta Hui said, "As soon as there's something considered important, it becomes a nest."[71]

Pleasant feelings condition the mind to desire and clinging, while unpleasant feelings condition dislike and aversion. Neutral feelings, on the other hand, can lead to delusion, or a lack of awareness of the true nature of things. This occurs because when we experience neutral feelings, we may become indifferent, complacent, or disengaged, preventing us from fully understanding the underlying reality of a situation.

Clopen pierces through the arbitrary duality of groupings to liberate the mind. As Alan Watts says, "The nonduality of the mind, in which it is no longer divided against itself, is *samadhi*, and because of the disappearance of that fruitless threshing around of the mind to grasp itself, *samadhi* is a state of profound peace."[72] In other words, clopen allows you to achieve *samadhi*, complete absorption, or a flow state (figure 129). It prevents you from clinging, which can lead you down common mental rabbit holes such as excessive worrying, imagining worst-case scenarios, or replaying past events and dwelling on what could have been different. Clopen breaks the need to classify and allows you to simply become aware of the here and now.

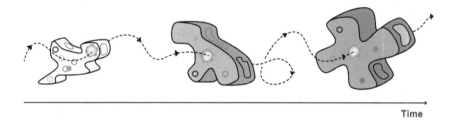

Figure 129. *Samadhi* is a flow state of pure concentration or absorption.

This book is littered with clopen examples. Most notably, your idea space can be clopen. Other examples include the young and old lady; the dress; the title *The Gateless Gate*; the mind is buddha and mind is not buddha; emptiness and everything; a topological singularity of infinite depth and zero measure; to me, here, I have no head, to you, *there*, I have a head; two people's experience of a sunset; spacetime is flat everywhere, yet curves; etc. Clopen phenomena can be found everywhere around us. What looks like one thing to one person may look completely different to another. As author Aldous Huxley said, "There are things known and there are things unknown, and in between are the doors of perception."

One effective approach to experiencing clopen is to explore the concept of emptiness. While it's not advisable to dwell in emptiness, it serves as a valuable tool for recognizing the nondual nature of reality. As with listening to the silence in between sounds, embracing the idea of nothingness can enhance our awareness of everything around us. In turn, observing everything around us can also reveal the nondual nature of reality.

The clopen nature of the world around us is not merely represented in our idea space; it is also manifested as a physical phenomenon in particle physics through the fascinating phenomena of *particle-wave duality*. Like light, the particles that make up an atom can either act as a particle or a wave, depending on how they are observed or measured. Understanding the duality of the physical world allows us to appreciate the nondual nature of reality on a deeper level by realizing all the matter that creates our world operates through a clopen phenomenon. The remainder of this chapter is dedicated to uncovering this phenomena, empowering you to perceive the nondual reality of nature and experience the state of *samadhi* wherever you go. In doing so, you can break free of the attachment of groupings and live a more peaceful life, ultimately lifting the next veil of illusion on our Path of Awakening.

PARTICLES, PARTICLES, AND MORE PARTICLES

Before we delve into the details of particle-wave duality, let's first take a closer look at the fundamental particles that make up our universe. Understanding the building blocks of matter not only allows us to appreciate the nonduality that exists at the core of our existence, but also guides us in discerning what the world is so we can better understand what the world is not, which is key to awakening.

What are you made of? From a regular distance away, you would probably say you're made up of skin, hair, bones, muscles, teeth, etc. Upon closer inspection, you'd change your answer to various types of cells or bacteria. If you look even closer, then you'd probably say you're made of molecules and atoms. But what actually is an atom?

The evolution of the atom is quite unique. A long time ago, the Greeks believed atoms were like tiny grain particles (figure 130-a). Then, physicist Ernest Rutherford created a model where the electrons orbited a nucleus, which consists of protons and neutrons (figure 130-b). Today, instead of orbits, the electron acts more as a "cloud" where it's impossible to pinpoint the exact location of the electron (figure 130-c).[73]

THE CLOPEN NATURE OF REALITY 181

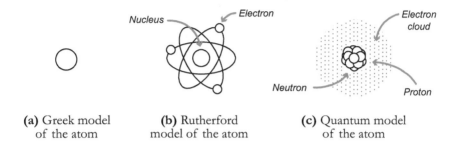

Figure 130. The evolution of the atom.

Digging further into the fundamental building blocks of matter, we can ask: *What are neutrons and protons made of?* To address this question, we look to the comprehensive classification of particles called the *Standard Model* (figure 131). Within this model, the three left columns represent all the matter in our universe, commonly called *fermions*, while the last two columns are the forces carrying particles called *bosons*. Each particle is characterized by its name, mass, and charge.

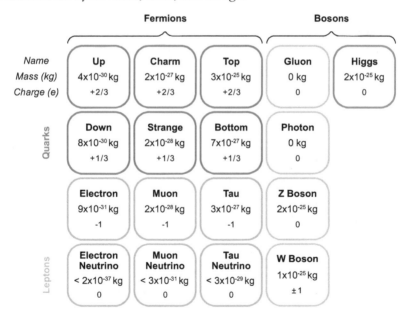

Figure 131. The Standard Model of Particle Physics.

Quarks, the first two rows of fermions, are subatomic particles that combine to make *hadrons*, like protons and neutrons. For example, a proton consists of two up quarks and one down quark (figure 132-a); and a neutron consists of one up quark and two down quarks (figure 132-b).* In both particles, the gluon is the force that holds the quarks together.

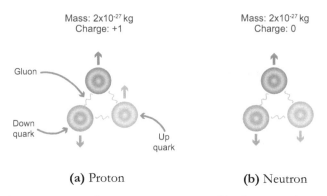

Figure 132. Protons and neutrons are types of hadrons made up of quarks.

The *leptons*, the last two rows of fermions, are the other matter particles. They have various functionalities and play a large role in radioactive decay. The most common lepton is the electron, which is usually found "orbiting" the nucleus of an atom. The electron, muon, and tau particles each have a neutrino counterpart. Neutrinos are often dubbed "ghost particles" as they seldom interact with ordinary matter.

Each fermion, or all of matter, has an associated mass and charge. Mass indicates weight. Particles with mass move slower than the speed of light, while massless particles move at the speed of light. Neutrinos are so tiny, no one has been able to detect their mass. It is suspected that they must not be larger than the values in the table.[74] Charge is the electrical charge. It is a physical property that causes a particle to experience a force when placed in an electromagnetic field, like in a wire. The charges here are relative to the charge of an electron (e), which is -1. For instance, a proton (two up quarks and one down quark) has a charge of +1, while a neutron (two down quarks and one up quark) has a charge of 0, or is neutral.

---

* A short introduction on quarks can be found in the supplemental material on the website.

Leptons and quarks make up all the matter in our universe. The bosons, or force-carrying particles, dictate how matter, or energy, interacts with other matter, or energy. Each of the four forces has its corresponding particle (figure 133). The *gluon* is the particle that mediates the strong force, which keeps the nucleus of atoms together by acting on quarks. The *photon* is the particle that mediates the electromagnetic force. It is more commonly known as "light" and acts on any particle with a charge, including other force-carrying particles. Both gluons and photons have zero mass, move at the speed of light, and have zero charge.

The Z and *W bosons* are particles that mediate the weak force, which acts on all matter. The weak force is responsible for the radioactive decay of atoms. For example, in large stars, a proton and electron can combine to form a neutron and electron neutrino. This reaction shrinks the radius of the atom from a size of 10^{-8} cm to 10^{-13} cm and causes a massive boom, known as a *supernova*. Note the W boson can have a charge of either ±1, while the Z boson is neutral.

The last particle is the *Higgs boson*, which gives mass to quarks, charged leptons, and the W and Z bosons. It is still a mystery how neutrinos obtain their mass, since it is uncertain whether they interact with the Higgs boson.[75]

At this point, you may be asking yourself: *What about gravity? Where is its force-carrying particle?* Great question! You may recall that gravity is essentially an illusion caused by the curvature of spacetime. For instance, if you place a bowling ball on a trampoline, then the trampoline will dip. If you then put marbles on the trampoline, they will tend to "gravitate" toward the bowling ball. Something similar happens in spacetime where matter or energy deforms spacetime. Then, the curvature of spacetime causes objects to "gravitate" toward one another.

The conundrum around finding gravity's force-carrying particle is a problem commonly known as *quantum gravity*. Since gravity is essentially a pseudo force that manifests itself as the curvature of spacetime, that's like asking: *What is the fabric of spacetime?* No one really knows the answer to this question. Therefore, gravity is often omitted from the Standard Model. That said, the leading theory is a hypothetical particle called the graviton, which interacts with every particle and moves at the speed of light, 186,000 miles per second.

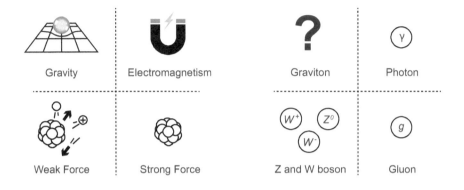

(a) The four forces of nature   (b) The force-carrying particles

Figure 133. The forces of nature and their corresponding force-carrying particles.

All the matter we're made up of only constitutes a small portion of the total universe. According to the European Space Agency's Planck Telescope, dark energy (responsible for the continuous expansion of space) makes up around ~68% of the universe, *dark matter* (gravitational seeds responsible for galaxy formation) makes up ~27%, while regular matter and force-carrying particles (all the particles in the Standard Model) make up the other ~5%.[76]

By delving into the intricacies of these particles and their behavior, we have set the stage to explore particle-wave duality. This knowledge, deeply rooted in the physical world, will serve as a catalyst for our journey toward a more profound appreciation of the true, nondual nature of reality.

PARTICLE-WAVE DUALITY

The defining trait of all the particles in the Standard Model is particle-wave duality, which allows a particle to exhibit both particle and wave-like behaviors depending on how precise the measurement is. In other words, contrary to the common perception that particles follow linear, straight trajectories, they can also display wave-like properties. Particle-wave duality is the most physical representation of clopen, as it governs the behavior of everything we're made of and interact with. To gain a deeper understanding of particle-wave duality, let's examine one of the most renowned experiments in physics:

THE CLOPEN NATURE OF REALITY

the *Double Slit Experiment*.

Picture a gun placed to the left of two walls. The first wall has two slits for the bullets to go through, while the second wall is a backstop with a line of bullet detectors (figure 134-a). The gun is hard to control, and every time you shoot it, bullets fly in all directions. Of course, as you shoot the gun, the bullet can ricochet off the slits and land in any particular spot. The goal is to see where the bullets land for three different experiments: (1) shooting with only the top hole open (figure 134-b), (2) shooting with only the bottom hole open (figure 134-c), and (3) shooting with both holes open (figure 134-d). In the third experiment, where both holes are open, the resulting probability distribution is simply the sum of the first two experiments, where only one hole was open at a time. Explicitly, the probability distributions, which represent the frequencies at which a bullet lands at a particular location, simply add up: $P_1 + P_2 = P_{12}$.

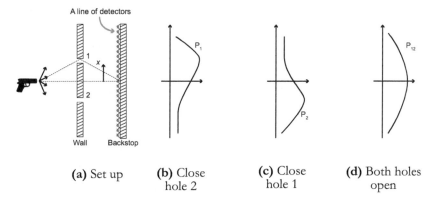

Figure 134. When performing the Double Slit Experiment with a gun, the probabilities add up as there is no interference.

Now, instead of a gun, let's do the same three experiments with water and place an object that bobs up and down to create circular waves. Moreover, instead of detecting bullets, our detectors pick up the amplitude, or intensity, I, of the wave (figure 135-a). The goal is to see how the intensities differ for each of three experiments: (1) only the top hole open (figure 135-b), (2) only the bottom hole open (figure 135-c), and (3) both holes open (figure 135-d). Unlike the gun example, the intensities no longer add up ($I_1 + I_2 \neq I_{12}$). The reason is simple: waves interfere with each other.

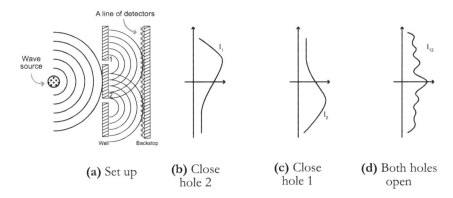

(a) Set up (b) Close hole 2 (c) Close hole 1 (d) Both holes open

Figure 135. When performing the Double Slit Experiment with water, the probabilities interfere with each other.

There are two main types of interference, creatively named: *constructive interference* and *destructive interference*. The names are exactly what you would expect them to be. Essentially, add the amplitudes of two waves to get the resultant wave. If the waves are "in synch," then you get constructive interference (figure 136-a). If they are "out of synch," then you get destructive interference (figure 136-b).

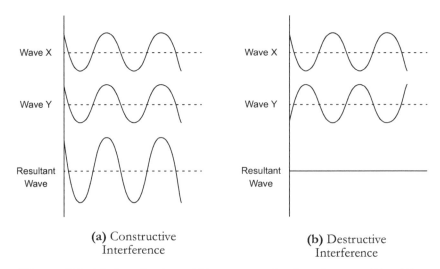

(a) Constructive Interference (b) Destructive Interference

Figure 136. **(a)** Amplitudes add in constructive interference, **(b)** while amplitudes cancel out in destructive interference.

THE CLOPEN NATURE OF REALITY

Now, what would happen if we sent a particle, like an electron, through the same process (figure 137-a)? Again, the goal is to see where the electrons end up for each of three experiments: (1) shooting with only the top hole open (figure 137-b), (2) shooting with only the bottom hole open (figure 137-c), and (3) shooting with both holes open (figure 137-d). In this instance, the particles do not seem to shoot in straight lines, like the gun. In fact, interference occurs and the electron behaves like water. This may seem contrary to our common perception that particles travels in a straight line.

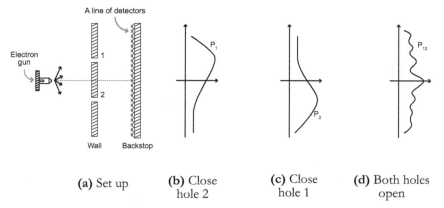

(a) Set up   (b) Close hole 2   (c) Close hole 1   (d) Both holes open

Figure 137. When performing the Double Slit Experiment with electrons, the electrons interfere with each other.

It gets weirder. This time, we're going to place a light source behind the slits that flashes every time it detects an electron going through one of the two holes (figure 138-a). That way, the light source will be able to tell us whether the electron has gone through hole one or hole two. One final time, we run the experiments to see where the electrons end up for each of three runs: (1) shooting with only the top hole open (figure 138-b), (2) shooting with only the bottom hole open (figure 138-c), and (3) shooting with both holes open (figure 138-d). This time, when both holes are open and we're able to see which hole the electron goes through, there is no interference, and the probability distributions simply add together!

By "watching" the electrons, we have changed their motion. When we do not watch which hole the electron goes in, the electron behaves like a wave, with interference. When we watch the electron carefully to see which hole it goes through, it behaves like a particle, without interference. This not only applies to electrons, but every single particle in our Standard Model!

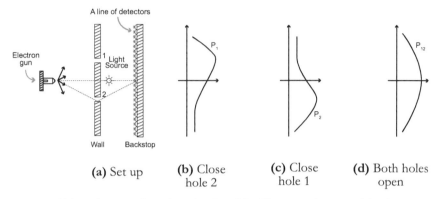

(a) Set up (b) Close hole 2 (c) Close hole 1 (d) Both holes open

Figure 138. When performing the Double Slit Experiment with electrons and a light source that detects which holes the electron goes through, the electrons do not interfere.

To try to understand the magic of particle-wave duality, we turn to clopen.* In terms of an analogy, open can represent a wave (figure 139-a), closed can represent a particle (figure 139-b), and clopen represents particle-wave duality (figure 139-c). When you are merely looking at a particle, it acts like a wave. Make a precise-enough measurement and it acts like a particle. At the end of the day, every particle has the ability to act either as a particle or as a wave. The situation is clopen.

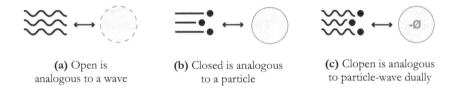

(a) Open is analogous to a wave (b) Closed is analogous to a particle (c) Clopen is analogous to particle-wave dually

Figure 139. Particle-wave duality is analogous to open, closed, and clopen.

---

* Quantum mechanics explains this phenomenon through the uncertainty principle. See supplemental material on www.TheIdeaSpace.io.

THE CLOPEN NATURE OF FREE WILL

As we have seen, the concept of clopen can be applied to various aspects of our reality, including the mysterious particle-wave duality. Another area where clopen might offer a unique perspective is the age-old debate on free will. The question arises: *Do we possess free will and the ability to make conscious decisions, or are our actions predetermined by a series of random events?* Interestingly, the concept of clopen suggests that we might simultaneously have and not have free will.

On one hand, when geneticist Robert Plomin describes DNA, the building block of cells, he states: "The single most important thing to know is that DNA consists of dumb molecules that blindly obey the laws of chemistry."[77] If quarks are dumb molecules that blindly obey the laws of physics; if DNA are dumb nucleotides that blindly obey the laws of chemistry; if cells are dumb organisms that blindly obey the laws of biology; then are we dumb humans that blindly obey the laws of the universe? If everything that creates us is a random assortment of goods, then would we not too be a random assortment of decisions? From this perspective, it would seem as if we are merely strangers with front-row seats to this crazy world we call "life" with no free will.

On the other hand, is it possible free will is the simple act of shining the light of mindfulness onto an idea space? Thus, when faced with choice points, is our free will to choose the future event we desire, based on our past and present, and let the world come to us? Then, if every moment of our lives is a choice point, does free will occur all the time?

In true Zen fashion, the situation is clopen. Life seems to strike a delicate balance between actively moving forward and being passively pulled forward by future events. This dynamic might be a reflection of the nature of the universe itself—the Big Bang propelling us forward, while the immense black hole at the center of our galaxy drawing us in. This duality of existence could explain why the philosopher Soren Kierkegaard once said, "Life can only be understood backwards; but must be lived forwards." In embracing this complex interplay, we can find wisdom in an ancient Taoist saying, which encourages us to harmoniously flow with life's currents: "Don't push the river."

THE CLOPEN WORLD

Overall, our idea space as a whole is clopen. Now, it would appear all particles, or everything we're made of and interact with, also follow a clopen phenomenon in particle-wave duality. Thus, we lift the next veil of illusion on our Path of Awakening (figure 140). Prior, there was only particles vs. wave. Now, we understand we live in a clopen universe of particle-wave duality. When we try to measure the precise location of a particle, it acts in a completely different manner. I guess that's Nature's way of protecting itself. It would now seem as if our minds and the observable universe are both clopen.

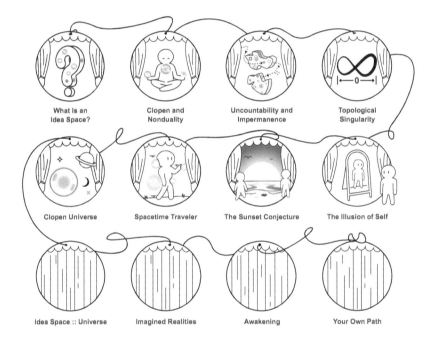

Figure 140. Understanding the clopen nature of reality lifts the veil of illusion on our Path of Awakening.

From an outsider's perspective, it is thought mindfulness only occurs when meditating internally. Of course, meditation is only the formal act of mindfulness. The practice calls us to take that mindful state into our day-to-day lives, as the comprehensive nature of mindfulness involves contemplating feelings and mind internally, externally, and both internally and externally. As guru Ajahn Chaa wisely puts it, "Some people think the longer you can sit, the wiser you must be. I have seen chickens sitting on their nests for days on end. Wisdom comes from being mindful at all times."[78] Or, as Zen master Huang Lung said, "Peaceful meditation does not require mountains and rivers: when you have extinguished the mind, fire itself is cool."[79]

In practicing internal mindfulness, we develop an awareness of our own thoughts and emotions, recognizing them without judgment or attachment. This allows us to maintain a balanced perspective and navigate through life with greater ease. To practice external mindfulness, one turns to inference and induction. It is at times obvious, either through a person's body language or words, whether they are feeling pleasant, unpleasant, or neutral emotions. Then, by inference of our own experiences of pleasant, unpleasant, or neutral emotions, we can know to some extent what the other person is feeling.

Joseph Goldstein does a great job summarizing why the practice of external mindfulness is important:

> Just as pleasant feelings condition desire, unpleasant ones condition aversion, and neutral feelings condition ignorance when we're unmindful internally, so too might seeing painful feelings in others trigger grief, sorrow, or denial in ourselves when we're unmindful externally.[80]

If we're not mindful of how others are, then their tinted idea space can overflow and tint ours. If someone else experiences unpleasant feelings, then this might cause pain or uneasiness within our idea space. If someone else experiences pleasant feelings, then this might cause jealousy or envy within our idea space. Thus, external mindfulness helps build empathy without letting ourselves be overcome by the others' emotions (figure 141).

Seeing the clopen phenomena of particle-wave duality gives you a new, external anchor point of mindfulness. It lifts the veil of illusion previously associated with duality. Before, there was only open vs. closed; particle vs. wave. Now, there's open and closed; particle and wave. Having an understanding that both your idea space and your observable universe are clopen allows you to remain mindful at all times, whether it's oriented internally or externally.

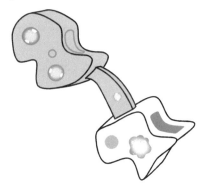

Figure 141. External mindfulness prevents someone else's idea space from overflowing into ours, while remaining empathetic.

In daily practice, this means constantly observing your thoughts and emotions, as well as those of others, without judgment or attachment. For example, when encountering a challenging situation, instead of immediately reacting, step back to see the space prior to the act of grouping. Observe the emotions and thoughts arising within you and others. If you catch your mind attaching to a particular thought or emotion, can you iteratively backtrack your thoughts to the original idea that caused the emotion (figure 142)? This approach helps you remain empathetic while maintaining your own emotional balance, allowing you to respond with clarity and wisdom in various situations.

Figure 142. Clopen allows you to break free from the attachment of groupings, thereby producing *samadhi*—a state of mind that is no longer divided against itself.

The clopen nature of reality can serve as a powerful tool for mindfulness, allowing us to navigate the complexities of life with insight and compassion. Clopen reveals a nondual approach to reality, which prevents you from attaching yourself to certain groups of thoughts. Plus, as we saw in Chapter 2, grouping is the most foundational aspect of the human experience; and, whenever you group, a duality arises. Therefore, seeing the clopen nature of reality prevents classifications and allows you to achieve *samadhi*, or flow, by producing a state where the mind is no longer divided against itself. As an Ancient once said, "Perceive before the act and you won't have to use the least bit of effort."[81]

Chapter 9

YOUR IDEA SPACE AS A REFLECTION OF YOUR UNIVERSE

"When you're learning the facts about life,
they're always about you."
- Dave Chappelle

Breathe . . . Simply breathe . . . There is nothing else here other than this moment . . . Sights . . . Sounds . . . Sensations . . . Thoughts . . . Notice the impermanence of this world and watch your elementary ideas flow in the wind . . . Stop for a moment, and take a few breaths for yourself . . .

Your observable universe is everything you see right now, and everything you're seeing right now is in the past. At the center of your observable universe lies your idea space of uncountable depth and zero measure. At the edge of your observable universe lies the creation of the universe, the Big Bang. In between lie trillions of galaxies, each with millions, billions, or even trillions of stars.[82] At the center of many of these galaxies lies a supermassive black hole partly responsible for the creation of its own galaxy. All in all, these elements provide the fabric for the large-scale structure of the universe.

After the Big Bang era, or our universe 380,000 years young, the universe consisted mostly of hydrogen. Other elements, like carbon, did not arrive until stars created them. Then, planets were able to form, which allowed for carbon life forms, like ourselves, to flourish. Thus, the words of Carl Sagan resonate quite beautifully: "The cosmos is within us. We are made of star stuff. We are a way for the cosmos to know itself." In other words, you are an aperture through which the universe looks at itself.

For centuries, humans have looked up at the night sky in search for answers and meaning in life. Does the universe have anything to teach us? Why are we here? What is the meaning of life? What is all this? While it is clear Carl Sagan's statement is certainly true, it is my belief the reciprocal is also true (figure 143):

The cosmos is a way for us to know ourselves.

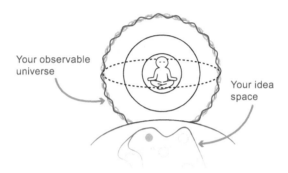

Figure 143. Your idea space is a reflection of your observable universe.

Explicitly, your idea space is a reflection of your observable universe. In other words, the large-scale structure of spacetime provides a great model and analogy for how the mind works. As an Ancient once said, "We are what we think. All that we are arises with our thoughts. With our thoughts, we make the world."

To see this more clearly, this chapter looks at the key similarities between your observable universe and your idea space. Namely, (a) how idea spaces form is analogous to (i.e., "::") how the universe forms, (b) how ideas form :: how stars form, (c) koans :: black holes, (d) galaxies :: principles, and (e) how idea spaces end :: how the universe ends.[*] Understanding the mind through this lens lifts the next veil of illusion on our Path of Awakening and allows you to further view your thoughts, sensations, emotions, and perceptions as objectively as you would view objects in the universe—without an "I."

---

[*] For a detailed analysis on the Big Bang, star formation, black holes, and galaxy formation, check out the bonus chapters at www.TheIdeaSpace.io.

THE BIG BANG

The first noteworthy similarity between your idea space and your observable universe is how they both form. Namely, no one knows how the universe started and no one knows where thoughts come from. Let's begin by exploring the mystery of the Big Bang, then see how it transfers to your idea space. As a disclaimer, no one theory of the Big Bang is definitely correct, since no one can confirm what actually happened at the beginning of the universe. These theories are more so ideas as to what *could* have happened, based on general relativity.

So, how did the universe start? What existed before the Big Bang? No one really knows. The first idea is nothing existed before our universe. Then, all of a sudden, a random "gravitational quantum fluctuation" caused the birth of our universe. Something came out of nothing. A *singularity* occurred. A singularity is a point where general relativity breaks down, which explains why no one knows exactly what happened. As physicist Stephen Hawking says, "The singularity is outside the scope of presently known laws."[83]

Overall, the main idea behind this singularity is an infinite, "space-like" surface. In other words, the Big Bang happened everywhere, all at once. BOOM. The key to a space-like surface is it creates a *particle horizon*.

To better understand this, let's look at an example and bring back our light cones and event p, our present moment in spacetime. Remember, the bottom edge of the light cone is your observable universe, which gets larger as it flows upward through time. On one hand (figure 144-a), we have a fictional universe created from a point. In this instance, you can see the history of all the particles that exist—all the particles have been observed by you. On the other hand (figure 144-b), the Big Bang is an infinite, space-like surface. You *haven't* seen the history of all the particles that exist. As you move up in time, you are continuously exposed to more and more particles. [84]

In truth, your universe acts like figure 144-b, with a particle horizon. The particle horizon represents the limit of the observable universe, the Big Bang, beyond which you cannot receive any information. The beauty of the particle horizon is stated in the following paragraphs.

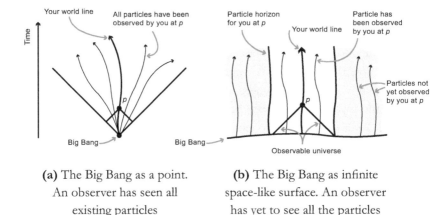

(a) The Big Bang as a point. An observer has seen all existing particles

(b) The Big Bang as infinite space-like surface. An observer has yet to see all the particles

Figure 144. The Big Bang singularity is hypothesized as an infinite space-like surface with a particle horizon.

Everything we see is in the past. Looking at the stars is like looking back in time; the light we see had to travel a long way to reach us. This means the edge of our observable universe is the Big Bang, the very beginning of time. So, the Big Bang didn't just happen, but is still happening everywhere, right now. In other words, the universe is always being created; and, as we saw through the Singularity Sunset, its creation impacts you uniquely.

Furthermore, as we saw in the fractals chapter, it is impossible to pinpoint exactly where we start and where we end. Thus, your Non-Self includes every unique layer of your entire observable universe, all the way to the Big Bang, the outermost layer.

This brings us to a beautiful insight: Your Non-Self is in a state of perpetual renewal, because your unique perspective of the Big Bang, your "Source", is constantly being created. In essence, as the universe continues to unfold, your Non-Self is continually rebirthed. This ongoing process gives you the opportunity to constantly reinvent yourself.

In this ever-evolving cosmos, your Non-Self is always a work in progress.

So, if the greater universe started out as this infinite, spacelike surface, then what did our observable universe look like at time zero, or at the Big Bang? Well, according to Stephen Hawking, our observable universe must have started with zero measure to evolve in a qualitatively similar manner. [85]Thus, a fantastic mental model for our presently observable universe as

YOUR IDEA SPACE AS A REFLECTION OF YOUR UNIVERSE   199

it was during the Big Bang is a topological singularity, which is represented as the uncountable points of Cantor Dust condensed into a single point (figure 145)! In other words, it is possible—though not definite—that our observable universe started out as nothing, ∅, then turned into something.

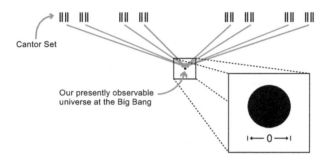

Figure 145. During the Big Bang, the region of spacetime, which eventually grew to become our presently observable universe, can be represented as a topological singularity.

The universe starting out as nothing then growing into something is one possibility (figure 146-a). Another possibility is the Big Bang wasn't actually a singularity, but instead a sort of "re-bound" from an old universe re-bounding to create ours (figure 146-b). In other words, another universe existed before ours that contracted to a small-enough size, skipped the singularity, and re-bounded to create our observed universe.[86] All in all, the only thing we know for sure is no one knows how the universe started.

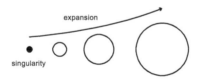

(a) A cosmological model where our universe starts as a singularity

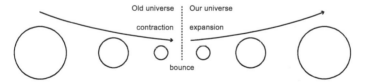

(b) A cosmological model where our universe starts as a bounce

Figure 146. Potential beginnings of our universe.

200 THE IDEA SPACE

After the Big Bang event, the rest of the Big Bang model comes into play (figure 147). More details can be found in the bonus chapters, but the key point is more changed in the first two minutes of our universe than the remaining 13.8 billion years! The Big Bang model ends 380,000 years after the start of our universe when the universe goes from being opaque to transparent during recombination. At this point, we're left with the famous Cosmic Microwave Background (CMB). Since everything we see is in the past, the CMB exists in our observable universe today as a relic of that time. Plus, because the CMB pervades the entire sky, it prevents us from directly observing events that occurred before it. In other words, the CMB sets a limit to how far back in time we can virtually see.

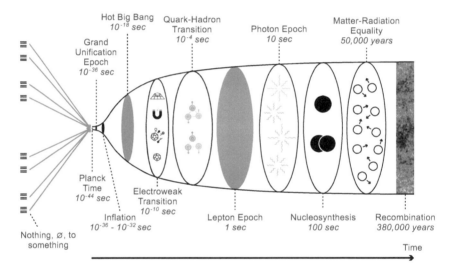

Figure 147. A mental model of the Big Bang era.

THE BIRTH OF AN IDEA SPACE

Where do thoughts come from? Where does an idea space come from? Just like the Big Bang, no one knows. For instance, does an idea space start from the empty abyss that is a topological singularity (figure 148-a)? Or does an idea space start as a bounce from one idea space to the next (figure 148-b)? Or is it a little bit of both: some start from nothing, while others start from an older idea space?

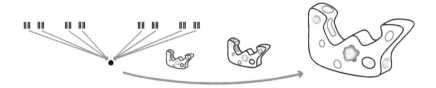

(a) An idea space forms from a singularity

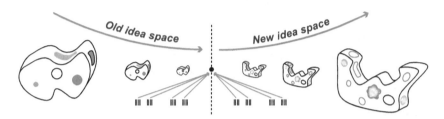

(b) An idea space forms from a bounce

Figure 148. How do idea spaces form?

Furthermore, as in the Big Bang, more changes occur in the creation of an idea space than the rest of the idea space itself. For instance, imagine something unpleasant happens to you. As soon as it happens, your idea space reacts with a thought that triggers an emotion. Although the thought itself may fade away, the emotion tends to tint the idea space for a long time, far beyond the lifespan of the thought alone. Before you know it, you're creating myriad fictitious scenarios in your head, filled with rage, recounting the event. In these moments, it's important to remember you're really just angry at a thought.[87]

Take the example of someone cutting you off in traffic: You might feel anger directed at the person who cut you off, but in reality, your anger stems from the thought that their action was disrespectful, dangerous, or inconsiderate. It's the interpretation of the event through your thoughts that evokes the emotion. Everyone's been there, including myself.

When we are unmindful, unpleasant experiences condition aversion and dislike, pleasant experiences condition desire and clinging, and neutral experiences condition disillusionment or apathy. Once you note an emotion, can you find the thought that triggered the emotion? From there, can you let go of any attachment to that thought? Once you do, you'll notice the half-life of negative emotions reduces greatly.

The goal of practicing mindfulness is to notice the arrival, the passing, and both the arrival and passing of mental formations. When we can see the complete materialization of an idea space, we can limit the stress certain thoughts impose on our minds and bodies, thereby allowing us to live with more peace and serenity.

STAR FORMATION

Star formation represents an interesting analogy to the formation of ideas within an idea space. Once again, we'll take a high-level view of our physical world first, then see how it relates to the idea space.

After the Big Bang era, hydrogen was the predominant element in the universe. Over time, these hydrogen molecules started to clump together via gravity to form giant gas clouds called *nebulas*. As hydrogen molecules got closer to one another, some collided, releasing heat in the form of an outward pressure, known as *resistance*, to combat the gravitational contraction (figure 149). For instance, imagine yourself getting pushed down in all directions. You wouldn't like that very much, would you? You'd probably try to fight against the collapse by pushing out. Well, that's exactly what stars do.

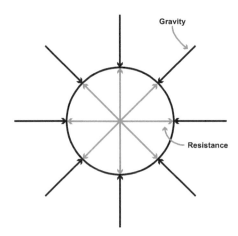

Figure 149. When the internal pressure caused by the collision of hydrogen atoms is as strong as the gravitational contraction, a protostar is born.

Eventually, when the internal resistance is equal to the gravitational contraction, a *protostar* is born. Around the protostar lies an *accretion disk*, which is simply a disk of gas that surrounds a large-enough stellar object. For a star like our sun, the protostar process takes around 10,000 years. Some tiny stars, known as *brown dwarfs*, only make it this far in life, then "die."

For larger stars, the process keeps going until *fusion* occurs. Fusion is the process by which smaller elements fuse to create larger elements, like hydrogen into helium. At this point, a star has reached its *main sequence*. All systems on! For a star like our sun, fusion occurs 100 million years after inception.[88]

From here, what happens depends on the size of the star. Small stars will continue fusion for billions of years and turn into a *Red Giant*. When small stars die, the outer layers gently fly into the cosmos to form a *planetary nebula*,* while the core turns into a *white dwarf*. A large star will continue fusion for millions of years and turn into a *Red Supergiant*. When large stars die, the outer layers bounce off the core of the star and a magical phenomenon known as a *supernova* occurs, leaving behind a *neutron star*. Supernovas are beautiful, because they light up the sky and are around 10^{10} (10 billion) times more luminous than our sun![89] For even larger stars, fusion also occurs for millions of years, but they "die" as a *black hole*. A black hole is a star so dense it rips through the fabric of spacetime.

The graphic below (figure 150) highlights the journey of stars where $M_\odot$ represents one solar mass, or the mass of our sun, 2×10^{30} kg. Stars get less massive as they live, shedding mass in the form of radiation, or energy, in accordance to Einstein's equation, $E = mc^2$. For example, our sun sheds around four million tons of mass every second. So, a small star starts at around, or below, 8 $M_\odot$. Large stars that turn into neutron stars usually start out around 8 to 20 $M_\odot$ in their early days and die at around 1.5 to 3 $M_\odot$.[90] Massive stars that turn into black holes are greater than 20 $M_\odot$ at birth and greater than 3 $M_\odot$ at "death".

---

* Don't be fooled by this misnomer—no planets are involved in creating these nebulas!

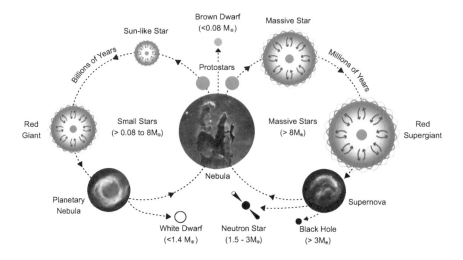

Figure 150. The complete life cycle of stars.[91]

FORMATION OF IDEAS

We are a way for the cosmos to know itself; the cosmos is a way for us to know ourselves. What does the life cycle of stars mean for our idea space?

To recap, stars have three paths of life, depending on their mass. First, all stars start out as protostars (figure 151-a). If a star fails its main sequence, then it turns into a brown dwarf (figure 151-b). If the star is large enough, then it will go throughout its billion-year lifespan and turn into a white dwarf (figure 151-c). If a star is massive, then it will last for millions of years and turn either into a neutron star or a black hole (figure 151-d).

It is my belief that ideas go through a similar trend except, instead of mass, the lifespan of an idea depends on its energy. *I thought mass and energy are the same by $E = mc^2$?* They are. Here, I don't mean the same energy found in the science of objects, things we can measure. I mean energy found in the science of the first person, things we can't measure. This energy usually comes in the form of *excitement*. In other words, how does an idea feel? Are new ideas flowing around it like an accretion disk would a star? Do those surrounding ideas fuel the initial idea? Is the idea able to grow and flourish? This is not something you can acutely measure. It is something you feel.

YOUR IDEA SPACE AS A REFLECTION OF YOUR UNIVERSE 205

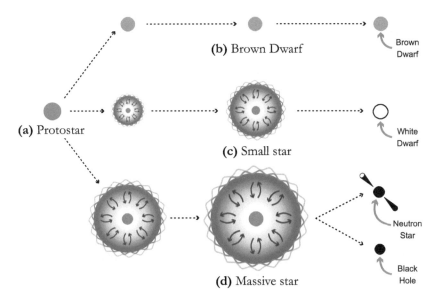

Figure 151. The three paths of a star.

To make this analogy explicit, ideas start out as *protoideas* (figure 152-a). In a sense, these are the thoughts, emotions, sensations, and perceptions left after a *brainstorm*. Like a protostar, protoideas don't take long to form. You're simply seeing what's out there. Then, depending on how much energy the idea has, it can take one of three paths. Sometimes, ideas suck. That's the fact of life. In these instances, the idea will turn into a *tiny idea* and end as a *poor idea* (figure 152-b). If the idea is good and picks up steam, then it'll ignite! It will go through its own main sequence and turn into a *small idea* (figure 152-c). If the idea is great, then it'll burn bright with so much energy and turn into a *massive idea* (figure 152-d). Of course, the lifespan of ideas is finite, so small ideas turn into *insights*, while massive ideas can turn into a *brainblast* or *koan*. A brainblast is similar to a supernova turning into a neutron star, while a koan is analogous to a black hole.

A couple of notes. First, everyone has poor ideas. To combat this, one turns to *stoicism*. One of the key tenets of stoicism is: Just because you have a thought, doesn't make you the owner of the thought. Did you provide the thought? Or did it simply arise?

That said, simply because you are not the owner of your thoughts, doesn't mean you have no responsibility for them. As the Buddha once said, "Bhikkhus, whatever a bhikkhu frequently thinks and ponders upon will become the inclination of the mind.'"*, 92 In other words, if you're constantly thinking about thoughts of sensual desire, then, over time, the mind will incline to thoughts of sensual desire. Thoughts carry momentum.

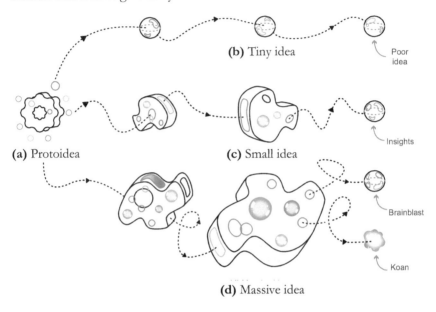

Figure 152. The different types of ideas.

Two great stoic exercises to combat poor ideas are *negative visualizations* and *the last-time meditation*. For negative visualization, simply think of a fictitious event that would cause unpleasantness to arise. Practice this now and take a second to think of something, like a piano falling on you from the sky. Don't linger on the idea for long. Is this event actually happening? No! Can you be grateful for the life you have now, at this very moment? Instead of wanting the life you don't have, like living on a yacht, want the life you do have, such as living in your current home. Experiment with this exercise a few more times throughout your day.

The last-time meditation involves imagining you are doing an activity for the last time. For instance, the last time reading. The last time walking. The last time eating. The last time seeing. The last time breathing. How

---

* A bhikkhu is someone who has taken up the path of enlightenment.

would your perspective change if this were truly the case? Would you not be eternally grateful for the current moment? Considering that life is impermanent, this is the last time you are living in this exact moment! Can you cherish this?

You could try a variation of the last-time meditation and practice a *first-time meditation*. Simply approach an activity as if it were the first time you were doing it. This helps promote beginner's mind and allows you to participate in an activity, even if it is your millionth time doing it, with fresh eyes. Finally, if you're feeling adventurous, imagine it is both the first and last time you're doing an activity. This mindset allows you to fully embrace the present.

Second, small ideas follow a similar path to small stars. Life evolved thanks to a small star, our sun, so even small ideas can have hidden beauty. Small ideas start with an ignition. A spark. This will bring forth much excitement. When this happens, you'll continuously feed energy into the idea and watch it grow at a slow, consistent pace. Upon the passing of the idea, you'll be left with insights. When you achieve insights, you develop discernment, which allows you to abandon ignorance. In other words, insights develop a calm, tranquil, clear knowing where the mind is free of desire, wanting, and restlessness.

Lastly, massive ideas follow a similar route to massive stars. They start with an ignition that fuels the idea with great haste, as a great deal of energy is gathered in a short amount of time. Massive ideas are the ultimate form of inspiration. And, inspiration is the purest form of ecstasy. There is a catch! As Naval Ravikant says, "Inspiration is perishable—act on it immediately." Since the massive idea gathers energy so quickly, it does not last very long. If you're able to act on it, then do so quickly!

The beauty of massive ideas is what's left after their passing. After awareness leaves the idea in question, the massive idea will turn into either a brainblast or an elusive koan. In both instances, there is a supernova: a moment of enlightenment, or an aha moment. Enlightenment is best summed up by the old expression:

> *Before attaining enlightenment, mountains are mountains, rivers are rivers.*
> *During enlightenment, mountains are no longer mountains, nor are rivers rivers.*
> *But after accomplishing enlightenment, mountains are mountains, rivers are rivers.*[93]

The key here is understanding the "mountains are mountains, rivers are rivers" experience *after* enlightenment is not at all the same as the "mountains are mountains, rivers are rivers" experience *before* such a realization. Although the moment of enlightenment lasts only for an instant, its remnants last for an eternity (figure 153). As an Ancient said, "When a single phrase is understood, you transcend a billion."

(a) Mountains and rivers

(b) The Koan Experience

(c) Mountains and rivers

Figure 153. Koans are the ultimate form of enlightenment.

Over time, it is important to remain curious and to investigate each idea. As Rick Rubin states, "If you have an idea you're excited about and you don't bring it to life, it's not uncommon for the idea to find its voice through another maker. The idea's time has come."[94] Ideas have immense potential and can lead to transformative experiences when we give them the attention and energy they deserve. By actively engaging with our ideas and nurturing them, we can cultivate a mindset that allows us to grow, learn, and ultimately contribute to a more profound understanding of ourselves and the world around us. When we ignore or do not properly foster our ideas, then the "Source" will find another way for the idea to manifest itself, even if it's through another person.

The arising and passing of a particular idea leaves a residue of that idea in your idea space, similarly to how stars leave a residue in spacetime when they die. In your idea space, the residue takes on the form of a poor idea, an insight, a brainblast, or a koan. In turn, these celestial ideas are responsible for building the large-scale structure of our idea space. If our universe has different types of stars and black holes continuously curving the fabric of spacetime, then our idea space has different types of ideas and koans continuously curving the fabric of our idea space (figure 154). Therefore, when we learn about the universe, we can actually use it as a tool to learn more about ourselves and our idea space. As comedian Dave Chappelle says, "When you're learning the facts about life, they're always about you."

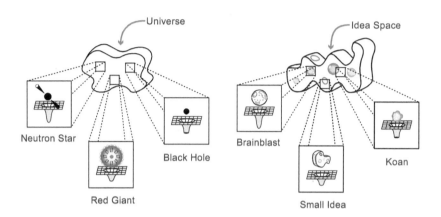

Figure 154. Ideas curve the fabric of an idea space, like massive stellar objects do the fabric of spacetime.

The best part about all these ideas is everything is recycled. For instance, within our universe, nothing goes to waste. Small stars emit planetary nebulas, which contain gasses that can be reused to create new stars. Small ideas emit *brainstorm dust*, which hold ideas that can be recycled for later use. Massive stars emit supernovas, which are responsible for creating 83 percent of the elements![95] Massive ideas lead to *enlightenment*, which is responsible for producing completely new, diverse ideas within your idea space. So, if you ever feel a moment of enlightenment, write down the associated thought! You never know when you may need it later in life.

Overall, similarly to how stars recycle themselves to nebulas, so too do ideas recycle themselves to form *brainstorms* (figure 155). Thus, the complete life cycle of ideas follows a similar trend to the life cycle of stars. Here, the words of Richard Feynman echo through the ether, "Ideas are always the most important thing, not the specific situation."

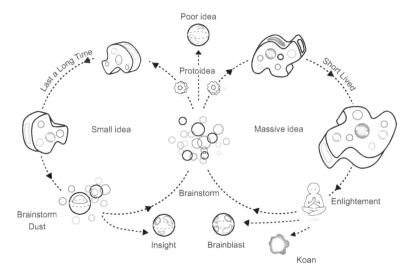

Figure 155. The life cycle of different types of ideas.

BLACK HOLES

Koans are analogous to a black hole. If a black hole is a rip through the fabric of spacetime, then a koan is a rip through the fabric of your idea space. Moreover, similarly to how a black hole completely distorts the reality around it, so too do koans. To understand this more closely, let's detail the properties of a black hole, then see how black holes relate to koans.

A black hole forms when a star is so dense, a singularity forms. Bring back to mind the example of a pen clicking through a piece of paper. Here the pen pushing down on the paper represents a star in spacetime, and the click represents the formation of the singularity. The black hole singularity has infinite density and curvature, yet zero volume.[96]

To protect itself from the outside world, the singularity develops a veil known as an *event horizon*. The size of this veil is known as the *Schwarzschild Radius*, and it is the maximum size a star can have before it collapses to form a singularity and black hole. At this point, the curvature of spacetime, or gravity, is so strong that even light itself cannot escape the star's grasp (figure 156). Any light emitted simply shoots back down toward the singularity.

For instance, the sun weighs 2×10^{30} kg and has a radius of 692,000 km. In order for the sun to turn into a black hole, it would need to collapse

YOUR IDEA SPACE AS A REFLECTION OF YOUR UNIVERSE 211

all of its mass into a radius of three kilometers. This won't happen since our sun is not massive enough and will likely die as a white dwarf. Another example: the Earth weighs 6×10^{24} kg and has a radius of 6,300 km. For Earth to turn into a black hole, it's mass would need to collapse into a radius of one centimeter, the size of your fingernail![97]

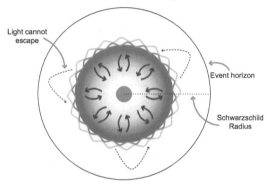

Figure 156. The Schwarzschild radius defines the event horizon of a star. Any signal sent from inside the event horizon will never be able to escape into the cosmos.

In our physical universe (figure 157-a), we are left looking at the hole's event horizon. We cannot see black holes themselves, because they emit no light. We can detect black holes because their gravitational effects are evident. For instance, in our embedding diagram (figure 157-b), the black hole rips the fabric of spacetime and curves the environment around it. Imagine placing a bowling ball so big on a trampoline that it rips the trampoline. However, instead of flinging back upward, the trampoline stays pulled down.

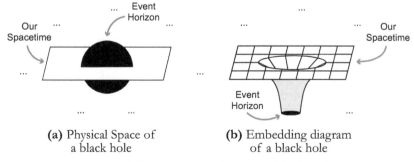

(a) Physical Space of a black hole

(b) Embedding diagram of a black hole

Figure 157. Representations of a black hole.

The complete gravitational collapse of a black hole is shown in figure 158. Here, a star implodes on itself until a singularity forms in the star. When this occurs, an *absolute horizon* appears and grows until it breaks through the surface of the star and creates the event horizon we're used to. The absolute horizon is the boundary between events that can send signals to the distant universe, like signals AA' and DD', and events that cannot send signals to the distant universe, like signals BB' and CC'.[98]

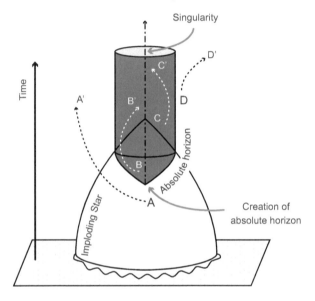

Figure 158. Embedding diagram of the gravitational collapse of a star turning into a black hole.

Most stellar objects in our universe are not standalone and have some sort of rotation, or spin, associated with them. For instance, the Earth rotates, which allows for day and night. The same principle holds true for black holes. When a star dies, it is usually spinning. Thus, the black hole it becomes also spins. After collapse, a black hole can continue to increase its spin if it is injected with fast-spinning matter. The maximum spin rate for a black hole is the speed of light.

When the black hole spins, it bulges out at the center, similarly to how the radius of the Earth's equator is 22 km longer than the radius to its poles (figure 159-a).[99] In our embedding diagram, we see the spinning black hole completely distorts the curvature around it (figure 159-b). Imagine placing a heavy, spinning bowling ball on a trampoline. As it breaks through the trampoline's surface, the ball continuously distorts and twists the fabric of the trampoline, thus forcing surrounding objects to orbit it.

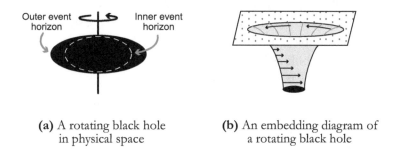

(a) A rotating black hole in physical space

(b) An embedding diagram of a rotating black hole

Figure 159. Representation of a spinning black hole.

Now, what happens if one were to fall through a black hole? Let's use a story developed by Kip Thorne to elaborate.[100] One day, Joe and Misty decide to use their imagination to travel to a collapsing star with some friendly neighborhood ants. Joe decides he will enjoy the gravitational collapse, while Misty watches from a safe distance. They've heard wacky things happen near black holes and therefore decide to remain in constant contact through a radio. Using these radios, Joe will send Misty continuous signals to let Misty know how much time Joe has experienced.

At the beginning (figure 160-a), Joe starts his journey with the ants and is able to emit radio waves at constant intervals to Misty. How much time Joe has experienced is illustrated in the bubbles. At this point, Joe's time and Misty's time are nearly identical.

Suddenly, the star Joe is on starts to collapse—the beginnings of a black hole (figure 160-b). When this happens, the curvature of spacetime starts to change. As the contraction occurs, Joe experiences a different time than Misty. For instance, when Joe sends the signal for "16 seconds," Misty's watch reads something more than 16 seconds. In other words, the spacing between the balls, which was at one point constant, changes. The balls are received by Misty at more and more widely spaced time intervals.

Eventually (figure 160-c), the star collapses completely and turns into a black hole when the star's mass fits inside its Schwarzschild Radius. Let's say this happens at 16 seconds in Joe's time. For Joe, his "16 seconds" signal is unable to escape the grasp of the collapse. The signal gets stuck with him and the ants. For Misty, she still continues to see a signal from Joe, but never sees the "16 seconds" signal. Instead, Misty sees Joe's signals asymptotically approach the "16 seconds" signal. First, 15.999 seconds. Then, 15.99999 seconds. Then, 15.99999999999 seconds...

In the end (figure 160-d), Misty will continue seeing the signal approach the 16 seconds bubble, but will never actually see the full 16 second bubble! The 16 second bubble will eventually get fainter and more red as Misty continues to wait for the signal until her final days. . .

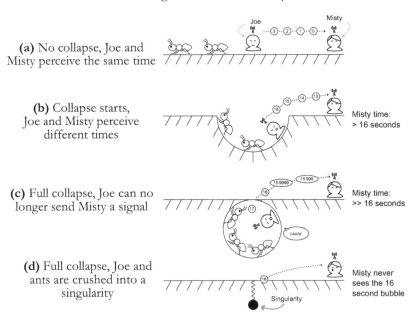

Figure 160. What happens during the gravitational collapse of a black hole.

This example does a great job of capturing what actually happens during the collapse of a black hole. Misty, at a safe distance away from the collapse, would never see Joe go through the black hole. Instead, Misty would see Joe and the 16 second signal frozen in time. Eventually, the signal would get fainter and "redder," but never fully disappear.

On the other hand, Joe's experience is quite different. As Joe and the ants continue their descent toward the singularity, they experience a random mix of gravitational waves that pull and contort their bodies in all

YOUR IDEA SPACE AS A REFLECTION OF YOUR UNIVERSE 215

sorts of directions (figure 161).* The closer they get to the singularity, the faster and stronger the oscillations become. No one know what happens as they approach the singularity as it is outside the known laws of physics. Does time cease to exist? Do space and time split up?[101] At this point the hypothetical quantum gravity would take over. I guess we'll have to jump into a black hole to find out . . . Any volunteers?

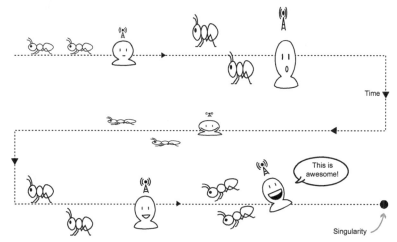

Figure 161. Gravitational waves distort the reality of Joe and the ants in all different shapes and forms until the singularity is reached.

Now, something else can happen when one jumps into a black hole. A *wormhole* can occur. Wormholes are theoretical play toys, but they are possible in general relativity. Wormholes are exactly what you expect them to be: they take you from one point in spacetime to another. This is similar to the concept of the re-bounding universe at the initiation of the Big Bang. Essentially, one could avoid hitting the singularity and be ejected into another point in spacetime, or another point in a different spacetime![102] If that's the case, then it is possible that Joe and the ants could be reunited with their dear friend, Misty (figure 162)! †

---

* This assumes a spinning black hole. In a stationary black hole, you would essentially be spaghettified: Pulled to infinity in one direction and then contract to nothing in the other two.
† Wormholes can occur if there is "exotic material" in the form of negative pressure density or when the charge density is larger than the matter density.

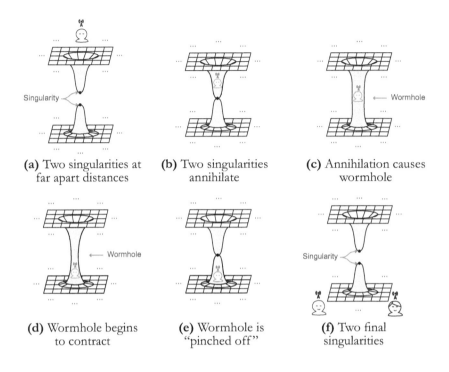

Figure 162. The creation process of a wormhole. Joe could pass through the wormhole and end up at a different point in the same, or different, spacetime.

KOANS: THE IDEA SPACE WORMHOLE

Koans rip through the fabric of your idea space and completely distort the idea space around it, like a spinning black hole would distort spacetime. Koans force other ideas to flow around them and serve as one of the foundational building blocks for the large-scale structure of your idea space, as they build your principles (see next section). For now, let's discuss how koans serve as a wormhole in between different idea spaces and within your idea space.

To start, it is clear no one can see your idea space, just as no one can see inside a black hole (figure 163). As Einstein said, "It is entirely possible that behind the perception of our senses, worlds are hidden of which we are unaware." Both singularities are masked by the veil of nothingness. We

YOUR IDEA SPACE AS A REFLECTION OF YOUR UNIVERSE 217

detect black holes through their gravitational effects. We detect idea spaces through words, body language, tone, or any other transmitter.

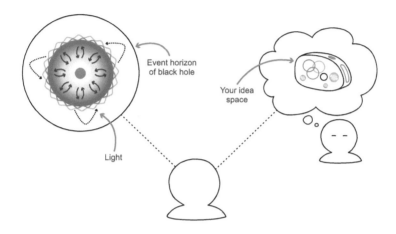

Figure 163. No one can see inside a black hole, and no one can see inside your idea space.

For these reasons, the mapping from one idea space to another idea space via spacetime is seldom perfect. For instance, although I have tried to capture the concept of an "idea space" in this book, the mapping of this concept from my idea space to your idea space via this book is not perfect (figure 164). There is a lot that has gone unsaid. Or, as Feynman says, "It is not so easy to say what we mean."

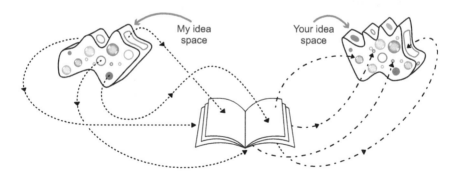

Figure 164. The mapping between idea spaces is seldom perfect.

Words only capture a portion of an idea space (figure 165-a). For example, how do you put an emotion, such as "love," into words? To combat this, we expand our list of transmitters to body language, pictures, math, tone, music, etc., to more closely approximate our idea space for others (figure 165-b).

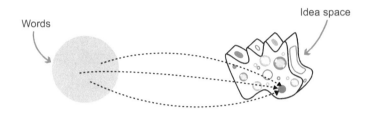

(a) Words capture a portion of an idea space

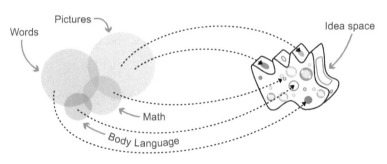

(b) Transmitters capture a larger portion of an idea space

Figure 165. Transmitters capture only a portion of an idea space.

Even with all these transmitters, the mapping is still not perfect. Of course, we can still convey what we're talking about, like the red apple from Chapter 1 or our mutual understanding of what an idea space is, but it's not the same as a perfect transmission from one idea space to the other. And, as author Khalil Gibran says, "Between what is said and not meant, and meant and not said, most of love is lost."

Here, koans come to the rescue. They serve as the ultimate form of information transfer in between two idea spaces. As an ancient Chinese proverb goes, "One showing is worth a hundred sayings." In a way, koans are a wormhole in between idea spaces (figure 166). They invite us back to a primordial experience that transcends space and time. They are a sudden moment of awakening. A sudden realization where mountains are no longer mountains and rivers are no longer rivers. A thought, emotions, song, or experience that perfectly captures the identity of an idea, or a moment in spacetime.

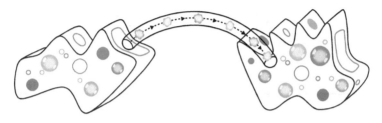

Figure 166. Koans act like a wormhole in between idea spaces.

For instance, a common idea space we subconsciously find ourselves in is: *Who am I? What am I? What am I going to be in life?* Answering these question seems to be the foundational layer of mostly everything we do. Thus, to capture the identity of this idea space, we turn to Zen master Dogen's Genjo Koan:

What is the way of the Buddha? It is to study the Self. What is the study of the Self? It is to forget the Self. To forget the Self is to be enlightened by all things.[103]

"I," your name, is simply one layer of your true Self. And, your true Self is your Non-Self. Enlightened by all things is to find beauty in all things. Perhaps you remember the Confucius quotation from the Mindfulness chapter: "Everything has beauty, but not everyone can see." Drop the "I" and attune to what is actually given. Sights. Sounds. Sensations. Perceptions. Sit in awe at the phenomenon of life. Can you find beauty in the smallest places?

Once the koan has captured the identity of an idea, the koan can be recalled by remembering the thought that triggered the koan. In other words, not only do koans serve as wormholes between two different idea spaces, but they also serve as wormholes within your own idea space (figure 167). For a specific example, what happens when you read: "No man steps into the same river twice, for it is not the same river and he is not the same man"? Does it conjure up an idea space of the past that captured the beauty of impermanence?

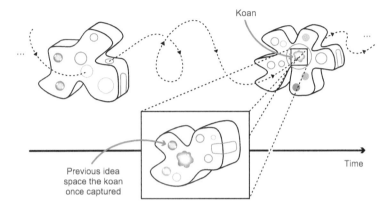

Figure 167. Koans serve as wormholes within your own idea space as well.

As you go throughout life, koans you come across can serve as mindfulness anchor points that allow you to traverse your own idea space. Like a video game, koans allow you to fast travel across your idea space. If you come across a difficult moment, refer to your koans, as they will shed clarity on the situation. For example, in regard to death, I love the koan by Graham Hancock: "Death is the beginning of the next great adventure." In this instance, the koan acts as a refresher of the truth, or the principle you built, to enable you to see the world with bare knowing or *yathabhutam*— just as it is. Seeing the world as it is brings great clarity to the situation. As you jump through the koan wormhole, it eases suffering by reducing the endless spiral of thoughts associated with the situation at hand.

Overall, koans are a clopen way to state the obvious. If the obvious never gets stated, then it ceases to become obvious. Therefore, the once obvious no longer becomes obvious. This leads to an interesting question: *If an idea is kept within an idea space, is it lost for an eternity?* Since an idea space is a topological singularity with zero measure (figure 168), the answer may be yes. For instance, if the "idea of an idea space" was kept in my head, would that information have disappeared forever, or would another spacetime traveler have come across the discovery? Moreover, if an obvious koan is never stated, then is that symbol of truth lost for an eternity?

YOUR IDEA SPACE AS A REFLECTION OF YOUR UNIVERSE

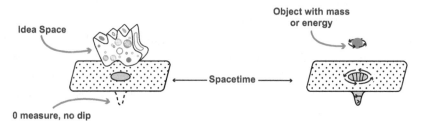

(a) An object with zero measure, like your idea space, has no direct effect on spacetime

(b) An object with measure has a direct effect on spacetime

Figure 168. Are ideas lost if they are not shared, since an idea space has no direct effect on spacetime?

GALAXY FORMATION

Your idea space has the potential to be filled with hundreds if not thousands of koans, just as the universe is home to trillions of galaxies. Both spacetime and your idea space are uncountably deep objects that house beautiful celestial creatures. In the universe, these come in the form of galaxies, which usually have a supermassive black hole at the center that helped create the galaxy. Analogously, in your idea space, supermassive koans are responsible for the creation of *principles*. All in all, black holes are responsible for large-scale structure of the observable universe, while koans are responsible for the large-scale structure of your idea space. To uncover this more thoroughly, let's briefly highlight the process of galaxy formation, then see how it relates to your idea space.

The key to galaxy formation is *dark matter*. No one knows what dark matter is, but it is hypothesized to be an undiscovered type of particle that mainly interacts through gravity, making it challenging to detect. Dark matter makes up ~27% of our universe, while regular matter (protons, quarks, molecules, electrons, etc.) make up ~5%. The remaining percentage is the mysterious dark energy, which is responsible for the continuous expansion of space.[104] The reason dark matter is difficult to work with is because it doesn't interact with light. So, like a black hole, we have to rely on its gravitational effects to detect it.

In terms of galaxy formation, dark matter serves as the gravitational seeds, usually called *dark matter halos*, that bring hot gas clouds together (figure 169-a). These hot clouds of gas then cool down to a certain temperature that allows for stars formation (figure 169-b). It is theorized the first generation of stars were much more massive than typical stars. Therefore, when they collapsed, they formed supermassive black holes. These black holes then formed a giant accretion disk, which eventually became the rest of the galaxy, like stars, planets, and other celestial objects. (figure 169-c).[105]

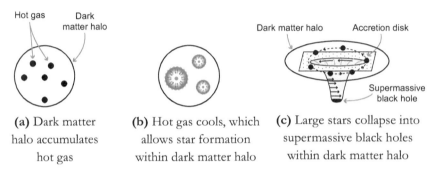

(a) Dark matter halo accumulates hot gas   **(b)** Hot gas cools, which allows star formation within dark matter halo   **(c)** Large stars collapse into supermassive black holes within dark matter halo

Figure 169. How an individual galaxy forms.

Galaxies can take on various shapes and forms. Three common types of galaxies are irregular, elliptical, and spiral galaxies. Our Milky Way is an example of a spiral galaxy. At the center of elliptical and spiral galaxies lies an *Active Galactic Nuclei* (AGN), a small area of the galaxy that shines extremely bright—so bright it can outshine the entire host galaxy, sometimes by as much as a factor of a thousand![106] For instance, an AGN known as *Quasar 3C273* shines 100 times more brightly than the brightest galaxies in the universe. The catch: a galaxy produces light in a region of around 100,000 light years in length, while 3C273 produces more light within a region of 1 light month![107]

A quasar is a supermassive black hole with a surrounding accretion disk that emits powerful jets in both directions, which can extend to distances of a million light years or more (figure 170-a)![108] For example, the Event Horizon Telescope team and the Hubble Telescope team were able to capture an image of the supermassive black hole at the center of Galaxy M87 (figure 170-b) and its powerful jet (figure 170-c).

YOUR IDEA SPACE AS A REFLECTION OF YOUR UNIVERSE 223

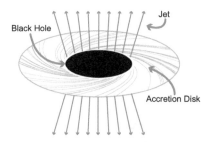

(a) Representation of a supermassive black home

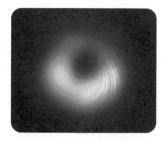

(b) Image of a black hole at the center of galaxy M87 produced by the Event Telescope team

(c) Image of galaxy M87 producing a jet by the Hubble Telescope team

Figure 170. A supermassive black hole accumulates an accretion disk and produces a powerful jet.

The universe started out 13.8 billion years ago with a bang. After 380,000 years, the universe consisted mostly of hydrogen. Stars began forming, which eventually turned into supermassive black holes, which created galaxies. The peak galaxy and star formation is known as *cosmic noon* and occurred around 10 billion years ago (figure 171). During this time, about half of all current stellar mass was formed.[109] Think of billions of spinning bowling balls absolutely tearing through the fabric of a trampoline. That's what's happening in spacetime around this time.

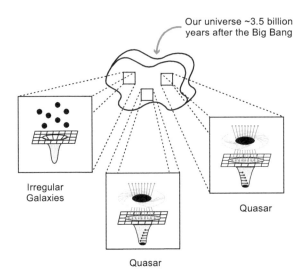

Figure 171. Cosmic Noon is the peak galaxy and star formation and occurred when our universe was ~3.5 billion years young.

PRINCIPLE FORMATION AND THE LARGE-SCALE CONVERGENCE OF IDEAS

Similarly to how the large-scale structure of the universe is dictated by black holes, so too is the large scale of your idea space dictated by powerful koans. In the universe, supermassive stars experience a supermassive supernova and turn into a supermassive black hole. In turn, new stars, planets, and other celestial objects are created around the supermassive black hole. In your idea space, supermassive ideas experience a supermassive moment of enlightenment and turn into a supermassive koan. As a result, the koan absolutely rips through the fabric of your idea space and forces all other ideas to flow around it.

The combination of a koan and the other ideas forced to flow around it is called a *principle*—a fundamental truth or proposition that serves as the foundation for your belief system, your behavior, and your chain of reasoning. A principle is analogous to a galaxy. It is an interconnected web of thoughts, emotions, sensations, and perceptions that revolve around the central koan, creating a coherent and consistent understanding of the world and guiding your actions and decisions.

For example, part of my idea space is listed below (figure 172). At the center of each subidea spaces of physics, math, wellness, and imagined realities lies a powerful koan, in the form of a quote, which shapes my principles, or thoughts, emotions, sensations, and perceptions, on the topic.

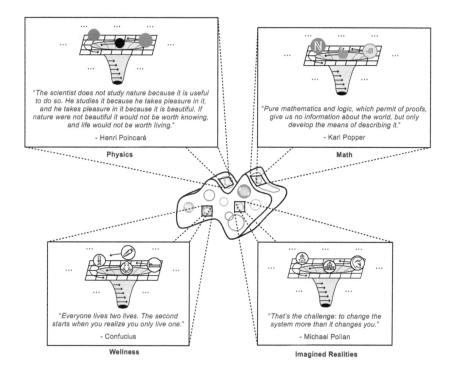

Figure 172. Principles in the author's idea space.

Mathematician Henri Poincaré's quote helps build a principle of curiosity and appreciation for the inherent beauty of the natural world. By recognizing the pursuit of scientific knowledge stems from an innate desire to understand the aesthetics of nature, it encourages a joy for discovery, a sense of wonder, and a drive to explore the world around me.

Philosopher Karl Popper's quote forms a principle of humility and awareness of the limitations of human knowledge. It reminds me that while mathematics and logic can provide a framework for understanding the world, they can never fully capture its true essence.

Sage Confucius's quote establishes a principle of self-awareness and presence in the moment. It encourages reflection on the finite nature of life and the need to make the most of the time I have. This principle influences the way I prioritize my actions and decisions, inspiring me to focus on what truly matters, pursue my passions, and live with intention and authenticity.

Author Michael Pollan's quote cultivates a principle of personal responsibility and critical thinking in a world often dictated by external forces and expectations. Instead of blindly accepting the status quo or following the crowd, it emphasizes the importance of questioning societal norms and making informed decisions based on my own values and beliefs. This principle guides me not only to cherish my own voice, but also to actively participate in shaping a more conscious and meaningful life, ultimately transcending the role of a passive observer.

Overall, my idea space is filled with principles like these, which are responsible for everything I do in life. As an Ancient once said, "If you penetrate in one place, you penetrate in a thousand." Something similar happens to you, but, of course, your idea space will look completely different than mine! That's why it's vital you write down your koans: They are responsible for the large-scale convergence of your ideas; they are responsible for building your principles.

Although enlightenment only lasts a moment, its effects linger for an eternity.* A koan is a rip in your idea space so powerful it forces all your thoughts, emotions, sensations, and perceptions to be influenced by it. Write down your koans! I cannot emphasize this enough. If there's one takeaway from this book, it's this: *Write down your koans!* Future you will thank present you because koans allow you to return to the moment of enlightenment. They allow you to return to your principles.

No one knows how these koans and principles form, they simply hit you like a brick wall. They're like a topological singularity: they look like nothing at first, then the whole world reveals itself. Perhaps their formation is due to the subconscious mind, in a similar way to how dark matter provides the gravitational seeds for galaxy formation (figure 173 a-b). Who knows?

---

* An item I didn't get the chance to touch on is black holes evaporate. So, koans evaporate too! The catch is black holes evaporate after 10^{67} years (10^{57} times the present age of the universe). Thus, koans essentially last a lifetime.

YOUR IDEA SPACE AS A REFLECTION OF YOUR UNIVERSE

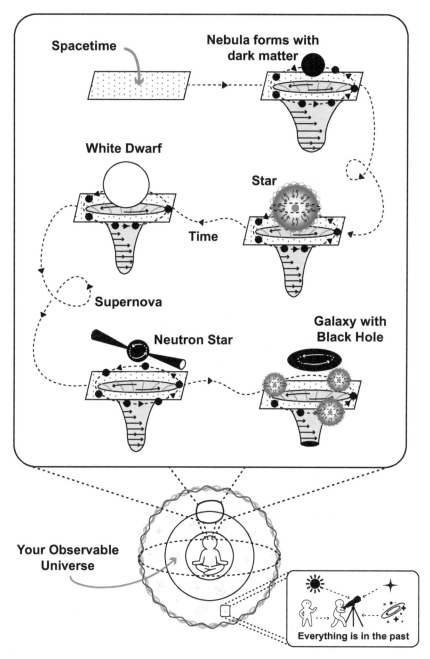

(a) Galaxy formation in your observable universe

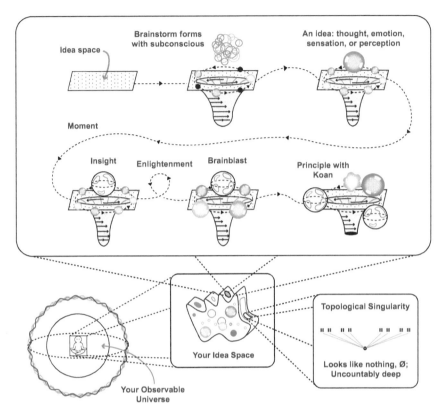

(b) Principles form simlar to the way galaxies form

Figure 173. Principles form similarly to the way galaxies form.

Practicing mindfulness is essential for recognizing and understanding the emergence of koans in your life, which subsequently form the principles that shape your beliefs and actions. In other words, on one hand, when we are not mindful, we miss the opportunity to see the formation of a koan, and we lose sight of a principle. Thus, principles do not form properly, and we become lost in the empty sea of our idea space. In this case, unwholesome actions arise, like killing, stealing, sexual misconduct, lying, harsh talk, backbiting, useless talk, covetousness, ill will, and wrong view of self.[110] In turn, we experience delusion, shamelessness, fearlessness of wrongdoing, and restlessness in the mind and body.

On the other hand, when we are mindful, the large-scale structure of the mind becomes clear as we notice the creation of koans and principles. A clear path flows from one principle to the next. In moments where we are unmindful (as mindfulness comes in short moments, many times), the principles guide us toward wholesome actions, like generosity, tolerance, tranquility, conscience, pliability, lightness of mind, balance, and proficiency. Thus, we can become a source of light not only for others, but also for ourselves.

Lastly, it would appear peak koan and principle formation occurs about a quarter of the way through one's life. This is our own version of cosmic noon. After the first quarter of life, one builds a mental schema of how the world works. In other words, people become set in their ways. To look past this veil of illusion, all that you, a spacetime adventurer, have to do is remain clopen to receiving koans as you endlessly fall through time. After all, the "Source" is always here constantly rebirthing your Non-Self.

THE END OF AN IDEA SPACE

How does an idea space end? Where do thoughts go once you notice them? No one really knows. The best we can do is relate it to how the universe will potentially end.

There are three main possibilities as to how the universe can end: Big Crunch, Heat Death, and Big Rip. Thankfully, before any of these scenarios actually happen, you'll probably be dead and the sun will have engulfed the Earth (around 7.5 billion years from now).[111] So, no need to worry. Keep living your life to the fullest.

One possibility is a Big Crunch, which is exactly what it sounds like. The universe expands to a certain point, then comes toppling back toward a singularity (figure 174-a). Of course, it is possible that right before hitting a singularity, the universe would bounce and create a new universe. Thus, each universe would get their own cycle of life and death. That said, while the Big Crunch is still possible, it doesn't seem super likely, considering the expansion of space has been accelerating for the last 4–5 billion years.[112]

Another possibility is Heat Death. While it would seem to imply that everything gets very hot and we all burn, that is not the case. Instead, Heat Death relies on the fact that everything (minus our small local group of galaxies) seems to be receding from us. Thus, at a certain point, everything outside our local group will have moved so far away, or outside the Hubble Radius, that everything will fade to black. The universe will be a very, very cold and dark place—hence Heat Death (figure 174-b).[113]

Lastly, Big Rip. As we talked about before, dark energy is something that is constant throughout space and time. If this were broken, if the energy density of dark energy were no longer constant but increased over time, then, at some point, every gravitationally bound system would be disassociated. For instance, galaxies would slowly break apart. Everything in the solar system would start slowly spiraling outward. The Earth would be pulled apart. Finally, atoms themselves would be torn apart (figure 174-c).[114]

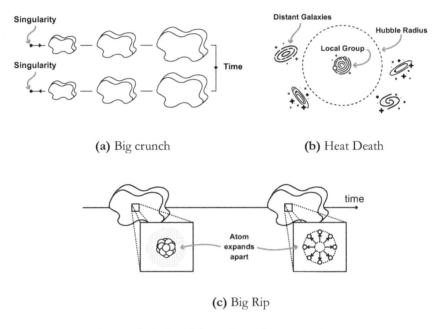

(a) Big crunch   (b) Heat Death

(c) Big Rip

Figure 174. Possible endings for our universe.

Now what about your idea space? Where do thoughts go once noticed? Who knows. Drawing inspiration from the universe, idea spaces can either crunch back into a topological singularity (figure 175-a), fly away into the subconscious leaving only the core idea space (figure 175-b), or maybe get pulled apart (figure 175-c).

YOUR IDEA SPACE AS A REFLECTION OF YOUR UNIVERSE 231

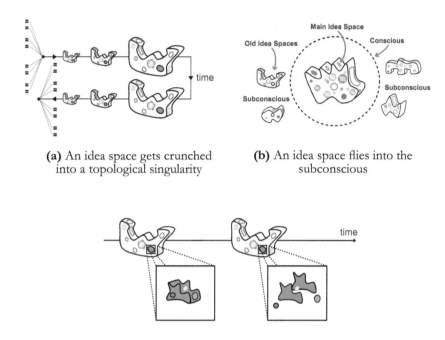

(a) An idea space gets crunched into a topological singularity

(b) An idea space flies into the subconscious

(c) An idea space gets pulled apart

Figure 175. Where do thoughts go once noticed? How does an idea space end?

The impermanence of the world dictates the constant arising and passing of idea spaces. Thankfully, the koan wormhole allows you to traverse in between idea spaces as they come and go. So, even though you may never fully understand where thoughts come from or where they go once noticed, you can rely on the principles you built to guide you down a path free from suffering, free from hatred, and free from ill will.

This is the power of your idea space. It lifts the next veil of illusion on our Path of Awakening; it allows you to detach from your Self, "I," or your identity, by viewing your thoughts, emotions, sensations, and perceptions as objectively as you would view the rest of the world (figure 176). Prior there was a division of "I," or "me," and the world. Now, you see that you are part of the world. You are Nature, and your idea space is merely a reflection of the world you see. As an Ancient once said, "We don't see things as they are, we see them as we are."

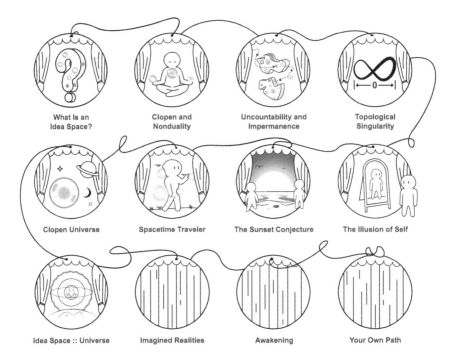

Figure 176. Seeing your idea space as a reflection of the universe lifts the next veil of illusion on our Path of Awakening.

Take a step back. Refrain from attaching to any particular idea, as clinging will only cause suffering. View the impermanence of the world both from the science of objects, what we can measure, and the science of the first person, what we can't measure. Be mindful of change both externally and internally.

Your idea space is who you are at zero measure. As close as others can get, they see nothing. But you are at this place of zero measure. Clearly, there is something there.

As you seek answers to traverse life, look to the stars. They have provided a guiding light for millennia to every form of our ancestors, and there is no reason their ancient wisdom will stop now. Everything you see is in the past, and the past is full of rich answers. All you have to do is *look*.

We are a way for the cosmos to know itself;
the cosmos are a way for us to know ourselves.

Chapter 10
IMAGINED REALITIES

"It is man who makes truth great,
not truth which makes man great."
- Confucian Principle

Let's try an experiment called one-minute *zazen*, or sitting meditation. With your eyes wide open, pick an object to stare at. It can be a book, a door, a wall, a chair, or even a desk. Now, stare at a fixed point on the object and do not allow your eyes to move. At the same time, bring your breath to a complete, or almost complete, standstill. Focus all your attention on the point. Prevent ideas from coming into your mind. Stare at that point. You may feel some thought-like actions come into your mind, but keep them under control. Focus all your attention on the point. Stare at the point. Do not let "perception" occur. If you have not started this exercise, look forward, pick a point, and try this one-minute zazen.[115]

It is estimated imagined realities started around 70,000 years ago with the rise of fictitious languages during the Cognitive Revolution.[116] Ever since then, humans have lived in two realities: *objective realities*, like trees, rivers, sights, sounds, thoughts, and emotions, and *imagined realities*, like words, money, laws, and governments. If awakening is to know what reality is not, then it's important to have an understanding of what imagined realities are, so we're not fooled or domesticated by them. Thus, we can lift our next veil of illusion on the Path of Awakening: *discerning objective reality from imagined reality*.

Imagined realities are a direct byproduct of macro idea spaces. If a personal idea space is the idea space of one individual, then a macro idea space is an idea space among multiple individuals. In a sense, a macro idea space creates a shared sense of reality for people and can be created through verbal or nonverbal interactions. For instance, a macro idea space can be created by talking to someone else, reading a book, or simply by ingesting information from another person.

Imagined realities are then formed when a subjective idea is shared among a mass group of people, thereby creating an *intersubjective reality*. An intersubjective reality is "real" because enough people believe it to be "real." If one person stops believing in the intersubjective reality, it still holds. For example, if one person stopped believing in the macro idea space of the "United States," it would not crumble (figure 177). It would take a majority of people to stop believing in the imagined reality for it to dissolve.

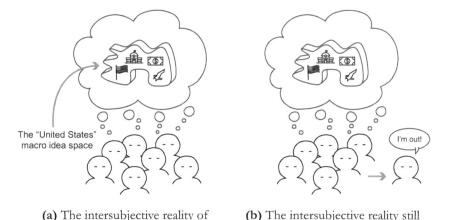

(a) The intersubjective reality of the United States

(b) The intersubjective reality still holds even if one person leaves

Figure 177. An intersubjective reality still exists even if one person stops believing in it.

More specifically, a macro idea space is an imagined reality, or intersubjective reality, if it fails the *Lindy Test*, or the test of time. The Lindy Test is a concept derived from the Lindy Effect, which was formalized by the father of fractal geometry, Benoit Mandelbrot. The Lindy Effect suggests that the future life expectancy of nonperishable things, like a technology or an idea, is proportional to their current age. Simply put, the longer an idea has been around, the more likely it will be around for a long period of time.

For example, a language is inherently not Lindy-proof, as its existence and relevance depend on the people who use it. Consider Sumerian, an ancient language spoken in Mesopotamia around 3,000 BCE. At one point, Sumerian was a thriving and widely used language. However, over time, it was replaced by other languages, and now it is considered a dead language, known only to scholars. Sumerian failed the test of time, thereby failing the Lindy Test.

Thus, a macro idea space is Lindy if it is good at *not dying*.[117] As Nassim Taleb writes, "That which is 'Lindy' is what ages in reverse, i.e., its life expectancy lengthens with time, conditional on survival . . . The longer an idea has been around without being falsified, the longer its future life expectancy."[118] Therefore, only the nonperishable can be Lindy. A good, Lindy-proof macro idea space is mathematics. For example, 1 + 1 = 2 was true millions of years ago and will be true millions of years from today.

This chapter is dedicated to identifying how the macro idea spaces of math, words, money, social norms, laws, religion, macropolitical systems, and scientific theories either pass or don't pass the Lindy Test. In turn, you'll build a better understanding of what an objective reality is and what an imagined reality is; thus, awakening. Of course, imagined realities are neither good nor bad. They simply are. In fact, most of them are quite useful, like words, but their usefulness should not be mistaken for reality.

MATHEMATICS

Does math pass the Lindy Test? Does math withstand the test of time? The short answer: yes. Math has been around for a long time and seems like something that will remain for years to come. This is because what was true in math yesterday will be true in math tomorrow. You may be wondering: *How long has math been around?* It's impossible to pinpoint a start, but a few examples help illuminate math's Lindy-proof nature.

The first example is the ant odometer. Ants, which developed around 100 million years ago after the rise of flowering plants, have exhibited mathematical behaviors. For instance, the Saharan desert ant, *Cataglyphis fortis*, travels immense distances to search for food. Each ant may travel as far as 160 ft until it encounters a dead insect before bringing it back to its hole, which is less than a millimeter in diameter.

Clifford Pickover summarizes an interesting experiment in his book, *The Math Book*, where it would seem as if ants know how to use math to count:

> By manipulating the leg lengths of ants to give them longer and shorter strides, a research team of German and Swiss scientist discovered that the ants 'count' steps to judge distance. For example, when ants had reached their destination, the legs were lengthened by adding stilts or shortened by partial amputation. The researchers then returned the ants so that the ants could start their journey back to the nest. Ants with the stilts traveled past the nest, while those with amputations did not reach it.[119]

Then, the same experiments were redone, except this time the ants started off with their modified legs. Here, the ants were able to compute the appropriate distances, suggesting that stride length is a crucial factor for ants as they "count" how many steps it takes to get from point A to point B.

The second example where "counting" can be found is with another social insect, the honeybee. As described in the wonderful documentary *Cosmos*, honeybees perform a *waggle dance* to inform other bees of where good plants or nests are located. As stated in the documentary, by performing this dance, bees can effectively communicate specific distances relative to the where the sun is. The "waggle" is done in the direction of the plant, relative to the light, and is maintained for however long the distance is (figure 178).

Figure 178. The waggle dance of the honey bee.

All this to say that counting, one of the underlying principles of math, has been around for as long as we can remember. Furthermore, some of the oldest math books, like Euclid's *Elements*, which was written in 300 BC, has content that still holds true today. Overall, it is clear math passes the Lindy Test and therefore is an objective truth. Simply put, what was true in math yesterday will remain true in math for the rest of eternity.

Thus, math becomes a useful tool for describing our world. As Karl Popper writes, "Pure mathematics and logic, which permit of proofs, give us no information about the world, but only develop the means of describing it."[120] The beauty exists when one takes a mathematical concept and applies it to a testable observation. This is the foundation of physics. If the mapping passes testing, then one has a powerful model used to approximate the world. As Galileo wrote, "The great book of Nature is written in mathematical language."

This is exactly what we did with the concept of an "idea space." We took the mathematical topics of the empty set, groups, countability, and measure to describe our idea space as a clopen grouping with zero measure and uncountable depth. Then, we subjected our idea space to various observations. Can an outside observer see the images in your mind? No, it has zero measure, so it looks like nothing. Can you count all your thoughts, emotions, sensations, and perceptions? No, your idea space is uncountable. In a way, math is the scaffolding we use to build not only our idea space, but also our reality (figure 179).

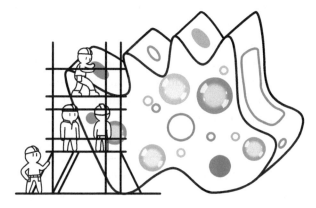

Figure 179. Math is the scaffolding we use to build our idea space & world.

WORDS

Words were made to transfer information from one idea space to another. It would not be startling if the passing of information from one entity to the next has been around since the dawn of time. For instance, when particles collide, information is transferred. In particle accelerators, such as the Large Hadron Collider, subatomic particles are accelerated to extremely high speeds and then made to collide with each other. The resulting collisions produce new particles, and the detection of these particles provides valuable information about the properties and behavior of the initial colliding particles, helping physicists better understand the fundamental forces and building blocks of the universe.

Therefore, the *idea of passing information* seems to pass the Lindy Test. In other words, transferring information has been around for a long time and will most likely stay for a long time. Conversely, words themselves do not seem to pass the Lindy Test. To understand this, simply look at the evolution of Indo-European languages over time (figure 180). Each language can be historically drawn back to other languages. It is evident the current language in use is bound to change, like it has for millennia. Will it be an evolution of English or another language that has yet to be invented? Who knows.

The main reason words evolve is because we need new ways to explain new phenomena. For instance, in this book, the words "idea space" and "topological singularity" were introduced. Neither of these phrases existed before, but now they mean something very precise. It's important to allow for the evolution of words to continuously enhance what we mean, whether it be in science or in the streets. Of course, in due time, these words will become obsolete as the next evolution of language comes into play.

So, words and languages are in a constant state of flux, and therefore do not pass the Lindy Test. The language I am using to write this book today is not the same language we'll be using 3,000 years from now. Words are merely squiggles on a page you understand, because you have a prebuilt disposition to them in your idea space. Just as Sumerian has gone functionally extinct, so too will English, French, Arabic, Mandarin, etc.

IMAGINED REALITIES

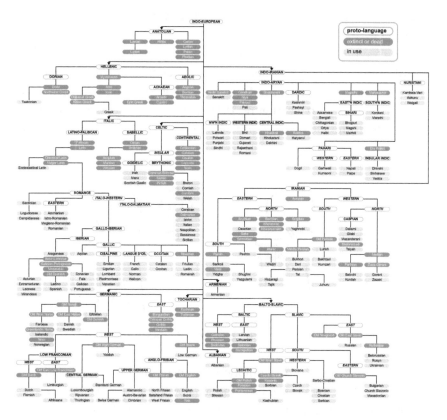

Figure 180. Classification of Indo-European languages. Dark gray are extinct languages. White are categories or unattested proto-languages. Light gray are languages still in use today.[121]

MONEY

Money. Barley. Guap. Moolah. Cheddar. Cheese. Coins. Cash. Dough. Benjamins. Paper. Loot. Copper. Dollars. Gold. Crypto. Bucks. Drachmae. Yen. Pesos. Euros. Bread. Dead Presidents. Stacks. Doge. Capital. Shekel. Shells. Wampum. Asset. Bitcoin. Resource. Check. Banknote. Quantum Crypto. These are some of the random ways money was, is, and will be represented.

Money, like words, is a transfer of information, except it also provides a reward or loss related to that information. Specific types of money, like the US dollar, the drachmae, or the yen, are not Lindy-proof. At one point, they didn't exist; then, they existed; and, there will come a time when they no longer exist.

That said, the concept of money as a transfer of information for a reward or loss may pass the Lindy Test. For example, the most elementary form of money comes in the form of trading one good or service for another good or service (figure 181). For instance, two species come together to form a *symbiotic relationship*: you scratch my back, I'll scratch yours.

Figure 181. Money is the transfer of goods and services.

A great example of a symbiotic relationship from *The Selfish Gene* by Richard Dawkins involves aphids and ants. Aphids are great at extracting the juice out of plants. After consuming the juice, aphids produce a liquid that is rich in sugar. In some cases, they produce more than their own body weight *every hour!* As these liquid droplets are excreted, several ants are right there to eat the droplets. The symbiotic relationship between the two species has become so profound that the ants actually "milk" the aphids to get their droplets. In certain instances, aphids will even hold back producing droplets until the ants are ready. Overall, the relationship benefits both parties: the ants receive sugar from the aphids, while the aphids receive protection from predators by the ants, ensuring their mutual survival and continuation of their symbiotic relationship.[122]

To simplify trade, *Homo sapiens* created a standard asset of money in the form of shells or skulls, which were often woven together to form necklaces. This was one of the first modern forms of money. The use of a common asset, like shells or skulls, made trading much more convenient compared to the cumbersome process of bartering different goods. In certain instance, tribes of *Homo sapiens* would meet in "fairs" in order to exchange goods for shells. That explains why seashells have been found up to 500 km from the nearest source.[123]

After shells, the first consistent form of money was barley, in use by around 3,000 BCE. Due to the shortcoming of its perishability, barley was quickly replaced by coinage around 2,500 BCE. Paper money, or banknotes, didn't make its way into our world until around 150 BCE. Nowadays, most of our money is transferred digitally. In the future, there is a chance a cryptocurrency, like Bitcoin or even quantum crypto, may eventually serve as a new form of money.

All in all, specific currencies do not pass the Lindy Test. The Drachmae, which was once used by the Greek empire, is no longer in use to today. It didn't exist; then, it existed; and now, it doesn't exist anymore. The same holds true for the dollar, yen, pound, Bitcoin, etc. While they may be forms of money *today*, they will most likely not be forms of money *tomorrow*.

The aspect of money that does seem to pass the Lindy Test is the ability to put value to our idea space through a reward and loss system. Throughout time, shells, barley, coinage, banknotes, and cryptocurrencies all have filled this void. In the future, quantum crypto or another source of money will certainly continue filling this void for as long as humans are alive (figure 182).

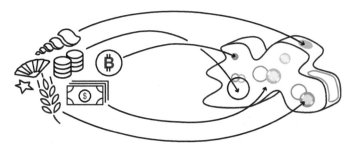

Figure 182. Money, in the form of shells, barley, coinage, paper, or crypto, assigns value to information in our idea space.

SOCIAL NORMS

If all your friends jumped off a bridge, would you? Probably not. If all your friends binge drank every weekend, would you? Maybe, maybe not.

Social norms, like table manners, are also an intersubjective reality that does not pass the Lindy Test. These norms exist and have influence only because we collectively believe them to be real. These types of imagined realities are especially deceiving, as they shape many of our desires, whether we want to acknowledge their impact or not.

Specifically, our desires can vary greatly depending on the specific intersubjective reality we are born into. For example, nowadays, many people desire to travel abroad for vacations. This concept would have been unimaginable to our chimpanzee ancestors, who would never have considered venturing into another chimpanzee band's territory for leisure. Likewise, ancient Egyptians were preoccupied with building pyramids and mummifying corpses, leaving no room for thoughts of shopping trips to Babylon or Phoenicia.[124] Romantic consumerism, a more recent phenomenon, will likely fade away just as the trend of pyramid-building eventually disappeared.

Simply because society or your friends are doing something does not mean you have to as well. Today, there is a large emphasis on going to college, getting a job, working nine to five till you are 60, then retiring. If this is the life you want to choose, then go for it. But remember, alternatives always exist.

Beware of groupthink. Do not accept something to be true simply because it represents the consensus of a group or because the media portrays it that way. Everyone is human and is capable of making mistakes. Make your own analysis of the situation. As Alan Watts writes:

> One must not forget the social context of Zen. It is primarily a way of liberation for those who have mastered the disciplines of social convention, of the conditioning of the individual by the group. Zen is a medicine for the ill effects of this conditioning for the mental paralysis and anxiety which come from excessive self-consciousness.[125]

Keep in mind, for something to not be true it only takes one example of it being false. Drawing inspiration from one of my koans: Simply because something is well known does not make it true; simply because something is not known does not make it false. What holds true for you here and now?

Although specific social norms are intersubjective realities that are not Lindy-proof, the concept of social norms themselves may pass the

Lindy Test. As it is clear, different species have different social norms. For instance, when a lone female moth emits a bouquet of pheromones to attract potential male mates, she is engaging in social behavior.[126] Without these unwritten rules, humans and other species would never be able to function in unison as well as they do.

Ultimately, your idea space interacting with the outside world and with other idea spaces within that world seems to be Lindy-proof. As such, it is important to see what social norms you engage with and tolerate—especially those that can be harmful or unskillful, like obsessively curating Instagram reels, relying on coffee to start your day, or adhering to a strict schedule of eating three meals a day.

What social norms are present in your life or in your group of friends? Can you be mindful of the culture these social norms create? Can you appreciate other people's social norms, even if they differ from yours? In the end, these norms are only real because we believe them to be real.

LAWS

The power of laws is not the actual law itself, but our ability and willingness to follow the law. Unsurprisingly, specific laws too are a form of imagined reality that are not Lindy-proof. Laws are a creation of our macro idea space. A great example of this can be found in the juxtaposition between the Code of Hammurabi (1,776 BC) and the American Declaration of Independence (AD 1,776). At a high level, both texts describe the rules of living for an organized group of people.[127]

The Code of Hammurabi was written in 1,776 BC in Babylon, which was then the world's largest city with more than a million subjects. The Babylonian Empire ruled most of Mesopotamia, which includes the bulk of modern Iraq and parts of present-day Syria and Iran. One of the most famous Babylonian kings today is Hammurabi, mostly because of the surviving text that bears his name, the Code of Hammurabi. This text served as a sort of uniform legal system for the Babylonian Empire (figure 183).

The piece of legislation starts off by stating it is an inspiration from the gods and then lists about 300 different "judgments," or laws. For instance, one judgment states: "If a superior man strikes a woman of superior class and thereby causes her to miscarry her fetus, he shall weigh and deliver 10 shekels of silver for her fetus." After all the judgments, Hammurabi declares these are "just decisions which Hammurabi, the able king, has established and thereby has directed the land along the course of truth and the correct way of life."

Figure 183. The Code of Hammurabi. It is written in the Akkadian (Semitic) language and is one of the original texts describing Sumerian law.[128]

This text was canonized and copied long after Hammurabi died and his empire lay in ruins. Therefore, it provides a good framework for understanding the Mesopotamians' ideals toward social order. This social order makes it clear that there is a set hierarchy of three classes—superior people, commoners, and slaves—and two genders. Furthermore, every class and gender has different monetary values associated with each type of judgment. The goal of the legislation was to provide rules for people to follow so the king's subjects could cooperate effectively. Their society could then prosper by producing enough food for its people, defend itself against enemies, and expand their territory to acquire more wealth and security.

About 3,500 years later, the Declaration of Independence was written when thirteen British colonies in North America felt that the king of England was treating them unfairly. On July 4, 1,776, the colonies declared independence from the royal crown. This declaration stated universal principles of justice, which, like those of Hammurabi, the authors said were inspired by a divine power. The American Declaration of Independence states as follows:

> We hold these truths to be self-evident, that all men are created equal, that they are endowed by their Creator with certain unalienable rights, that among these are life, liberty, and the pursuit of happiness.

The similarities between this document and the Code of Hammurabi are fairly evident. Both state certain principles, or laws, for the people under rule to follow, thereby allowing millions of humans to cooperate effectively and live peacefully. Furthermore, both are not documents of time and place as both were accepted by future generations as well. Even at public schools today, 250 years later, students stand for the national anthem and the Pledge of Allegiance.

The two texts present an obvious dilemma. Both the Code of Hammurabi and the American Declaration of Independence claim to outline certain universal principles, or laws—but those principles contradict each other. According to Americans, all people are equal. According to the Babylonians, people are unequal. The Americans would say the Babylonians are wrong, while the Babylonians would say the Americans are wrong. Yuval Harari, author of *Sapiens*, summarizes this well:

> Hammurabi and the American Founding Fathers alike imagined a reality governed by universal and immutable principles of justice, such as equality or hierarchy. Yet the only place where such universal principles exist is in the fertile imagination of Sapiens, and in the myths they invent and tell one another. These principles, [or laws], have no objective validity.

As is evident from these two examples, laws themselves are always changing. What was a law yesterday may not be the law tomorrow. Laws are imagined realities that do not pass the Lindy Test. That said, it does not mean these laws are not useful! Without them, the standard of living would be nowhere near where it is today. Yet, it is still important to distinguish reality from perceived reality.

RELIGION

Of course, the Code of Hammurabi and the Declaration of Independence represent only one type of law. Another type of law exists in religion, which provides a guiding light to spacetime travelers trying to make sense of this crazy world. Although there are many types of religions, the main one we'll focus on here is Zen Buddhism, since many of its principles have been used to develop this book.

In short, Zen is the religion of no religion. It requires you to study your own mind and make decisions accordingly. The principles of Zen are captured by the *dhamma* (Pali) or *dharma* (Sanskrit), which roughly translates to "the truth," "the law," "the teachings of the Buddha," or "the Buddha's method."[129] Essentially, these are the laws one would abide to achieve enlightenment or awakening according to Buddhist Nature.

Many of the principles we covered align with the Buddhist teachings, such as *anicca* (Pali), or impermanence, and *anattā* (Pali), or absence of Self. However, it is important to remember these Buddhist principles, too, are simply an intersubjective reality or imagined reality, existing and holding meaning only because we collectively believe in them. That said, these teachings are incredibly useful for navigating your idea space and finding a genuine, sincere, and harmonious purpose in life.

A nice summary of the *dharma* can found in the Buddhist Four Noble Truths, which build on top of one another. The First Noble Truth is that of *dukkha* (Pali), which roughly translates to "suffering" or "frustration":

> Now this, bhikkhus, is the noble truth of dukkha: birth is dukkha, aging is dukkha, illness is dukkha, death is dukkha; union with what is displeasing is dukkha; separation from what is pleasing is dukkha; not to get what one wants is dukkha; in brief, the five aggregates subject to clinging are suffering.*

The passage does not imply the whole world is *dukkha*, or suffering. As the last sentence states, these objects are only suffering should we cling to any of the five aggregates of thoughts, emotions, sensations, perceptions, or consciousness. Since the five aggregates are impermanent in nature, clinging onto an impermanent object is bound to create *dukkha*. In other words, if we cling onto any one part of our idea space, then we'll suffer (figure 184). The more we grasp at the world, the more it changes.

For instance, consider you're enjoying a delightful ice cream cone on a hot day. You're savoring the taste, the chill, and the aroma—experiences of the five aggregates. But then, it starts melting. Suddenly, your perfect treat becomes a sticky mess, and disappointment sets in. This is dukkha, the result of clinging onto the fleeting nature of the ice cream experience.

---

* Bhikkhu is anyone on the path to enlightenment, like yourself.

IMAGINED REALITIES

Figure 184. The First Noble Truth: clinging to our idea space causes *dukkha*.

This brings us nicely to the Second Noble Truth, which is the cause of *dukkha*:

> Now this, bhikkhus, is the noble truth of the origin of dukkha: it is craving, [or desire]; that is craving for sensual pleasures, craving for becoming, craving for disbecoming.

Craving for sensual pleasures means desire for sights, tastes, smells, and sounds, which are agreeable or pleasant. Craving for becoming means desire for a renewed existence, while craving for disbecoming means desire for nonexistence. In essence, desire, or craving, represents an agreement you make with yourself to remain discontented until your wants are fulfilled. In every instance, there exists a Self, or "I," responsible for the craving: "I want to go to the beach," "I want to be rich," or "I want this experience to end." In all situations, this "I" distracts your Non-Self from the present moment. In terms of idea spaces, the Second Noble Truth relates to the constant chase of another thought, emotion, sensation, or perception in hopes they are fulfilling (figure 185).

Figure 185. The Second Noble Truth: the constant chase for another idea space is the cause of *dukkha*.

To end this craving, the Buddha introduces the Third Noble Truth, or the end of *dukkha:*

> Now this, bhikkhus, is the noble truth of the cessation of dukkha: the remainderless fading and cessation, renunciation, relinquish-ment, release, and letting go of that very craving.

To end desire, understanding the impermanent reality of our world comes to the rescue. Craving is associated with a Self. Through impermanence we see there is no Self. The person we call "I" is always changing from moment to moment as your idea space and the universe are both uncountable. Since there is no Self, we can let go of these cravings associated with "I" and return to the present moment. We can live in the here and now to watch the ever-changing landscape of sensations, thoughts, emotions, and perceptions instead of getting caught in the imagined worlds our mind creates, which is the cause of *dukkha*. Simply put, want what you already have; don't want what you can't get. Live in your idea space of today (figure 186).

Figure 186. The Third Noble Truth: to end *dukka*, live in your idea space of today, instead of the idea space of yesterday or tomorrow.

Of course, ending, or renouncing, all craving is a nontrivial task. It is a lifelong journey encapsulated by the Fourth Noble Truth, which is the path leading to the end of *dukkha*, or the Eightfold Path (figure 187):

Now this, bhikkhus, is the noble truth of the way leading to the cessation of dukkha: it is this noble eightfold path; that is, right view, right thought, right speech, right action, right livelihood, right effort, right mindfulness, right concentration.

Figure 187. The Fourth Noble Truth is the path leading to the end of *dukkha*, or the Eightfold Path.

The first two are considered the wisdom group; the next three are considered the morality group; the last three are considered the concentration group. Right view encompasses all the others and implies seeing the world *yathabhutam*, or just as it is, with clear awareness. Right thought involves understanding that although we are not our thoughts, whatever we frequently think and ponder becomes the inclination of the mind.

Right speech is broken down into speaking truthfully, avoiding slander and gossip, remaining mindful of emotional tones, listening mindfully, and limiting useless talk. As an Ancient once said, "Immeasurably great people are turned about in the stream of speech."[130] Right action is the mindful action of avoiding unskillful acts, which are given as the five precepts: (1) no harm in thought, word, or deed to any living creature, including yourself, (2) avoid stealing and act on spontaneous acts of generosity, (3) only speak about who is present, (4) abstain from sexual misconduct, and (5) avoid heedless intoxication. Right livelihood is the ability to find a profession that is spiritually fulfilling.

Right effort is putting energy into wholesome states of mind, like compassion and *mettā*, or loving-kindness, instead of unwholesome state of minds, like thoughts of sensual desire, ill will, or cruelty. Right mindfulness involves not getting carried away by thoughts nor associating ourselves as those thoughts. Instead, mindfulness means returning to the present moment and seeing the world with bare attention: thoughts as thoughts, emotions as emotions, sensations as sensations, and perceptions as perceptions. In other words, mindfulness is seeing your idea space objectively in between topological singularities (short moments, many times). Right concentration is the ability to build momentum of mindfulness, or *samadhi*, either through one-pointed focused awareness or choiceless awareness.

Like any other law, the teachings of the Buddha themselves are imagined realities that do not pass the Lindy Test. Zen master Lin-chi captures this sentiment well: "Outside the mind there is no Dharma, and inside also there is nothing to be grasped... Just be ordinary and nothing special. Relieve your bowels, pass water, put on your clothes, and eat your food. When you're tired, go and lie down. Ignorant people may laugh at me, but the wise will understand."[131] In other words, when you perform an action, do so with full intent instead of getting distracted by other aspects of life.

Even the Buddha, in the *Diamond Cutter* scriptures, mentions you should not cling onto the gold dust of the *dharma*. Although gold (*dharma*) is precious, gold chains still bind and gold dust still blinds. In other words, the Path of Awakening is an imagined reality. It is not real. It is a tool used to live a life free from suffering. Thus it is said, "The Buddhas have not appeared in the world, nor is there any *nirvana*. They manifest such things as expedient means to rescue sentient beings."[132] That is why it is important to make your own *dharma*. Or, as the Ancients consistently state, "Don't die in the words of ancients."[133]

All in all, at one point, the laws of the Buddha, or *dharma*, didn't exist. Then, the *dharma* came into existence. Today, these teachings are constantly evolving due to the impermanent nature of reality. Most notably, Zen is the brainchild of Mahayana Buddhism mixed with a little Taoism and Confucianism. Buddhism arose in India, traveled to China where it became Ch'an, and finally found its way to Japan where it became Zen. In turn, an idea space is the brainchild of Zen and physics (figure 188). It is a union of the humanities and exact sciences formulated to describe the science of the first person using tools from the science of objects.

Religion, or the philosophy of life, is in a constant state of flux. Therefore, religions, like Buddhism whose core teaching is the *dharma*, are not Lindy-proof. Once again, this is not to say religious teachings, like Christianity, Islam, Judaism, Hinduism, and others, are not useful. In these teachings lie the ineffable, hidden truths of the science of the first person, the science that builds the principles of your idea space. As with the exercises at the beginning of this book (e.g., Müller-Lyer illusion), it is important not to take these religious teachings nor the concepts found in *The Idea Space* at face value. Instead, like any good scientist, investigate the teachings for yourself to see whether they line up with your perception of reality.* If something hurts to hear, then look for the truth in it.

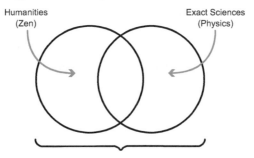

Figure 188. The Idea Space is a union of the humanities and exact sciences.

MACROPOLITICAL SYSTEMS

Macropolitical systems are imagined realities, which include any large organization of people under the umbrella of an intersubjective reality, like governments, companies, religions, or any other institution. Macropolitical systems do not pass the Lindy Test, as the macropolitical system of yesterday will not necessarily be the macropolitical system of tomorrow. For instance, the Persian Empire, which dominated for centuries, is no longer here today.

Governments are especially interesting since they are the driving force behind many of the imagined realities we've covered thus far. For example, the United States of America (USA) is a form of government. Under the USA, English is adopted as the main language, the dollar is used as the main form of currency, and the constitution is the law. Other countries have different governments with different languages, currencies, and laws. In a way, a government is the backbone of all other imagined realities.

---

* For me, I am religious to my Unknowns Unknowns.

Throughout time, governments have taken on many different shapes and forms, starting as small *tribes*, then growing to larger *chiefdoms*, and eventually becoming so large that states formed.[134] Independent of size, the main purpose of these macropolitical systems is *protectionism*: to protect the weak from being bullied by the strong.[135] For instance, the smallest tribe consists of a family whose family members are supposed to protect one another. Chiefdoms are supposed to feed their people and keep them safe from physical harm. States are supposed to protect their people both from physical and economical harm.

As the size of macropolitical systems grew, it became more difficult to organize everyone. Thus, new, unspoken social contracts were continuously adopted between the citizens of a macropolitical system and the main form of governance. Understanding these hidden social contracts is no easy task. As Karl Popper states,

> Social life is so complicated that few men, or none at all, could judge a blueprint for social engineering on the grand scale; whether it be practicable; whether it would result in a real improvement; what kind of suffering it may involve; and what may be the means for its realization.[136]

In this passage, Popper compares our understanding of a social world to a mechanical engineer's blueprint. With a blueprint, it is more trivial to predict the outcome of a design.* The same cannot be done when configuring social groups. One cannot predict what a certain group of people will do based on a law for the simple reason people do not act rationally! As Isaac Newton said, "I can calculate the motion of heavenly bodies, but not the madness of people."

Macropolitical systems, such as governments and empires, are constantly in a state of change and are therefore not Lindy-proof. For instance, at one point, the Roman Empire didn't exist; then, it existed; finally, it collapsed. David Whyte captures the impermanent nature of this well: "Whether we finally manage to get to Italy or not, whether we walk the evening streets of the imperial city or not, all of us will surely, if we live long enough to gain the perspective, see Rome before we die."[137]

---

* An engineer might counter, "Even blueprint designs seldom go as planned!"

This quote serves as a reminder of the transitory nature of human-made systems and the importance of embracing change as a natural part of life.

The ever-changing cycles of empires make sense since they are creations of an impermanent macro idea space. That said, an underlying feature across these macro idea spaces lies in the concept of open societies and closed societies. This is similar to the concept of open and closed idea spaces applied on a macro level.

A closed society is a macropolitical system that wants to go back to the way things were. It is effective tribalism. A closed society believes the ruling intersubjective reality of the time is more important than the individuals of that intersubjective reality. Namely, a closed society believes it is the purpose of the individuals to maintain and strengthen the state, or dominant macro idea space of the time. Nazi Germany, in 1933, is an example of a closed society, because its regime maintained strict control over its citizens, limiting their freedoms and imposing its ideology on all aspects of society.

An open society believes it is the purpose of the state, or intersubjective reality, to protect its individuals. Again, protection means preventing the weak from being bullied by the strong both physically and economically, a case echoed by both the weak and the strong. It need not mean self-protection too. Many people virtuously insure their own lives to protect others. In this society, individuals are confronted with personal decisions and there is no closure to the outside world. In an open society, members strive to rise socially. France, after the French Revolution in 1789, is an example of an open society, because it established a series of constitutions, which declared the aim of society to be the welfare and happiness of the people.

A good juxtaposition between these two ideologies can be found during the Peloponnesian War. The war was fought between Athens and Sparta after the disparate Greek city-states had assembled together for the first time to fight the Persian Empire, in around 478 BC.* After defeating the Persians, Athens thought to build a democratic empire by taming the seas to bolster trade. To achieve this, the Athenians decided to build "Long Walls" to connect inland cities to port cities though walled, protected corridors.

This philosophy did not sit well with the other powerful city-state of the time, Sparta, as they valued a more tribal lifestyle. Specifically, Spartans did not believe in trade with outsiders, nor humanitarianism (no individualistic

---

* Think the movie *300* (2006).

ideologies), nor universalism (Sparta vs. the world). Instead, they believed in mastery (slaves), tribalism (shut out foreign influence), and not becoming too large. One can see how this Spartan line of thinking directly contradicts the Athenian values of the time. Athens was turning into a blossoming democratic system that would be bolstered by free trade, while Sparta wanted to limit exposure to the outside world.[138] In a way, Spartans wanted to maintain the Spartan state through whatever means necessary (closed macro idea space), while Athenians wanted to improve the lifestyle of the people living in Athens (open idea space) (Figure 189).

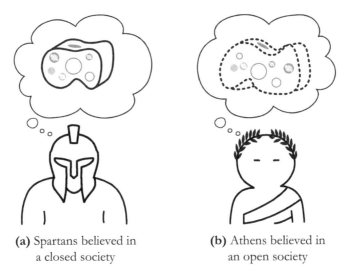

(a) Spartans believed in a closed society

(b) Athens believed in an open society

Figure 189. Closed societies believe the state is more important than its people. Open societies believe people are more important than the state.

Although many macropolitical systems are associated with division, macro idea spaces have the uncanny ability to bring people together. For instance, during the Dark Ages (AD 500 AD – AD 1,000), Europe consisted of hundreds of mini-kingdoms, each with their own castles, lords, knights, barons, and serfs. As the book *The Sovereign Individual* jokes, for years "cities and countries changed sovereignty the way that antique shops changed owners."[139] Times were dark, as illiteracy, blight, famine, and violence ravaged the land. This time period was a hodgepodge of disparate groups who were constantly fighting, as they all believed in a different macro idea space (figure 190-a).

To combat the violence, the Church started a movement called "The Peace of God," which recognized the overlordship of armed knights in exchange for a cessation of violence. The unification of knights and their followers into armies allowed the Church to organize the Inquisition, which lasted from the twelfth century to the fifteenth century. On top of providing support to poor farmers and curbing violence, the Church did a lot to improve productivity by increasing literacy and providing sponsored universities and education. For example, the Church created better buildings and robust architecture that allowed for more efficient trade and commerce. Overall, the new belief in the intersubjective reality of the Church allowed for the creation of a unified, open macro idea space (figure 190-b).

(a) Disparate kingdoms all believed in their own, closed macro idea space

(b) The Church unified an open macro idea space for all the disparate kingdoms

Figure 190. The Church macro idea space was able to unify the disparate kingdoms who each believed in their own macro idea space.

From about AD 1,000 to AD 1,500, the Church became the dominant form of governance and took hold of some common practices held by our government today, like law, record deeds, registering marriages, probating wills, licensing trades, titling land, and stipulating terms and conditions of commerce.[140] But, as author Matt Ridley states, "Empires, indeed governments generally, tend to be good things at first and bad things the longer they last."[141] Over the years, there seemed to be more churches, convents, monasteries, friaries, confessors, preacherships, cathedral chapters, endowed chantries, relic cults, religious co-fraternities, religious festivals, and new holy days. Services, prayers, and hymns grew longer and more complicated. Every day, new orders appeared to beg for alms.[142]

In turn, the once open macro idea space of the Church, which once helped set up schools and decrease violence, grew into a closed macro idea space filled with vast restrictions (figure 191). For instance, sexual relations between spouses were illegal on Sundays, Wednesdays, and Fridays. In addition, people could not have sex for forty days prior to Easter and Christmas. That adds up to no sex for 55 percent of the year! Furthermore, Pope Sixtus IV (pope from 1471 to 1484), who allegedly caught syphilis from one of his many mistresses, became the first pope to issue licenses to prostitutes and to levy a tax on their earnings.[143] Lastly, the Church enforced Canon Law on its alum mines, fisheries, and textiles monopolies to reinforce pricing by preventing customers from buying these assets from cheaper countries.[144] These vast restrictions made the general public hate the clergy and look for a new way of life.[145]

Figure 191. Over time, the imagined reality of the Church went from an open macro idea space to a closed macro idea space.

The key to replacing an intersubjective reality is to give people a new one to believe in. Through technological advances made available by the industrial revolution, like the printing press and gunpowder, the Church's reign dwindled as rich monarchies took over. During the 1,500s, fighting was no longer done in the name of the Church or whichever kingdom you were fighting for. Instead, it was fought in the name of economics.[146] With the introduction of gunpowder to the West, the largest army wasn't necessarily the one that won. Gunpowder provided a cheap and efficient form of violence. Now, monarchs with the most money started winning. Thus, the rise of monarchs gave rise to today's ruling intersubjective reality, the *nation-state*.

IMAGINED REALITIES

States in general have been around for six thousand years, but before 1,800 AD, they only accounted for a small fraction of the world's sovereignties.[147] The first, modern nation-state can be traced back to either the United States (1,776) or France (1,789). The rise of the nation-states is riddled with violence, as both the US and France started with bloody rebellions. Since then, many wars have been fought among them, including Napoleonic Wars (1,803-1,815), the Taiping Rebellion (1,850-1,864), World War I (1,914-1,918), and World War II (1,939-1,945). That's a lot of deaths, all in the name of an imagined reality.

After the world wars, the two dominant macro idea spaces were *mass democracies* and *socialist states*, or communism (figure 192). After all the wars throughout history, dozens of contending systems of sovereignty, like absolute monarchies, tribal enclaves, prince-bishoprics, direct rule by the pope, sultanates, and city-states, had come and gone over the past five centuries.[148] Why were these the final two on the podium?

Historian Charles Tilly provides a concise answer to the question, "States having the largest coercive means tended to win wars; efficiency (the ratio of output to input) came second to effectiveness (total output)."[149] How communist states were able to maximize coercion is clear: the government owns everything up front, then reallocates accordingly. In a mass democracy, citizens are allowed to own property and accumulate wealth, which is then taxed. Using this process, the mass democracy was able to place an even greater number of resources in the hands of the state compared to socialist states.[150]

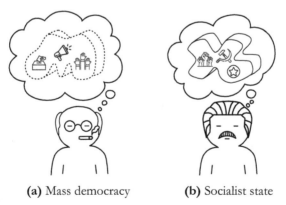

(a) Mass democracy (b) Socialist state

Figure 192. After WWII, the two dominant macro idea spaces were mass democracies and socialist states.

From this analysis, it is clear that specific forms of government do not pass the Lindy Test. Countries and governments, like the United States, China, France, Cameroon, Australia, Chile, Brazil, Morocco, India, Iran, and Germany, are only real because we believe them to be real. There will come a day when their existence vanishes. While specific macropolitical systems do not pass the test of time, it seems as if macropolitical systems in general may. Independent of group size, there seem to be imaginary rules in place to organize the cooperation of individuals. In small groups, this takes on the form of social norms. In larger groups, it takes on the forms of countries and laws. In other species, like ants and bees, it takes on the form of monarchies.

As we've seen, intersubjective realities are replaced when the people are given a new one to believe in. Latin was replaced by the Romance languages of French, Spanish, Italian, Portuguese, and Romanian. Gold and silver were replaced by the US dollar. The constitutional monarchy of France was replaced by *Declaration of the Rights of Man and of the Citizen*. Governments and macropolitical systems are in a constant state of flux (figure 193). During their birth, they usually come in the form of open societies, but at death, they usually take on the form of closed societies.

Figure 193. The impermanent reality of macropolitical systems.

Today, we are at the birth of a new macro idea space: the *Information Age*. One of the key drivers of large-scale changes is technology. The rise of micro-processing and computing has transformed our world as never before. For instance, the automotive industry was once the pride and joy of the Industrial Age. Nowadays, a software and hardware company like Tesla dominates innovation in the automotive industry.

The Information Age gives rise to the *sovereign individual*. In the new cybereconomy, it doesn't matter whether you are ugly, fat, old, disabled, white, black, Asian, Hispanic, Jewish, or Muslim. If you are a clear thinker, the opportunities for wealth creation will be unlike any we have seen before.

The Information Age will allow more equal opportunities for billions of humans in different parts of the world to thrive, as the spectrum of possibilities has been unlocked thanks to the internet.

In this new age, we no longer ask ourselves: *Who should lead?* As we've seen countless times throughout history, it is impossible to organize a perfect social structure. This leads to the centralization by a few, which turns into a closed society.[151] Instead, echoing the words of Karl Popper, it is best to work toward a piecemeal society where we accept that we are not masters of social engineering. Here, we work in an agile manner to try new approaches and change them if they don't work; and, if there are old obsolete rules, we trash them. Maybe, we even put a time limit on laws to see if they are worth renewing.

Mistakes will happen in piecemeal society. As they do, we must remember Isaac Asimov's wisdom: "To err is human, and to correct gently is humane." Thus, we can move past even an open society to a *clopen society*. Instead of asking who should rule, we ask, echoing the words of Popper once more, "How can we organize political institutions in a way that bad leaders can be prevented from doing too much damage?"[152]

The rise of cryptocurrencies and generative artificial intelligence (AI) will lead to an interesting answer to this question. One of the major shifts accompanying the rise of each of the macropolitical systems has involved money and technology. The Greeks used the drachma. The Church had various types of metals, and it was broken by the printing press. Most nation-states today have their own currency. How will the world change with a decentralized form of currency? If currencies are decentralized, will it be possible to build a more decentralized form of government? Furthermore, will the impact of AI be the modern day equivalent of the printing press? This is the unpredictable, yet beautiful, future we are responsible for building.

The world is truly headed for a great place. I am excited to live in this day and age where the rise of the sovereign individual is made possible by the Information Age. At the end of the day, it may be hard to know what to think politically. Here, I appreciate Nassim Taleb's fractal take:

> I am, at the Federal level, a libertarian; at the state level, Republican; at the local level, Democrat; and, at the friends and family level, a socialist. If that saying doesn't convince you of the fatouousness of left vs. right labels, nothing will.

SCIENTIFIC THEORIES

Specific scientific theories are not Lindy-proof, but science itself is. The simplest example of this is the evolution of the atom. As we saw, at one point the Greeks thought it was a tiny grain particle (figure 194-a). Then, physicist Ernest Rutherford created a model in which the electrons orbited a nucleus of the atom (figure 194-b). In recent history, we believed the nucleus consisted of neutrons and protons and the electron acted more as a "cloud," where it's impossible to pinpoint its exact location (figure 194-c).[153]

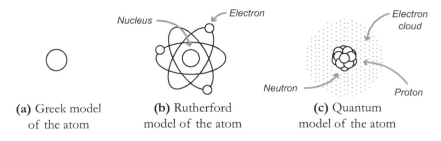

Figure 194. The evolution of the atom.

Today, this theory keeps evolving as the universe is viewed as a field of particles (figure 195). The next evolution of field theory seems to involve all of us living in a world of strings, yet there is no direct experimental evidence for string theory. Scientific theories are in a constant state of flux. Although quantum mechanics is the gold standard of today, that may change tomorrow. As physicist David Griffiths writes, "It is entirely possible that future generations will look back, from the vantage point of a more sophisticated theory, and wonder how we could have been so gullible."[154] Imagine realizing the world was round after millennia of thinking it was flat.

The reason scientific theories are constantly changing is because they are by default creations of our idea space. A scientific theory is a model we use to describe the world around us. It is half intersubjectivity of words and half objective reality of math. Stars are stars only because we named them stars. To our ancestors, they had a completely different meaning. To our future generations, they will probably have another completely different meaning. Of course, we can explain how a star works from a theoretical perspective, but the theory itself never fully encapsulates the full reality of nature.

IMAGINED REALITIES

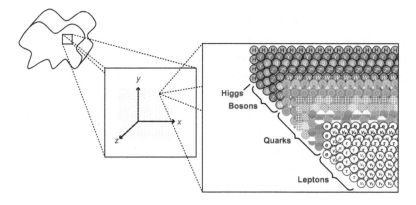

Figure 195. The universe modeled as an ever-changing field of particles.

Thus, a scientific theory is an idea that sits in our idea space, overlaid on top of reality (figure 196). Simply because a theory sits within an idea space does not mean it's not useful! Scientific theories have allowed us to accomplish amazing things in this world and should be continued to be taught to every child in the world, along with mindfulness!

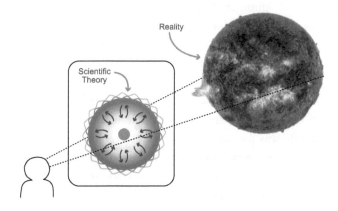

Figure 196. A scientific theory is an idea living in an idea space overlaid on top of reality.

Scientific theories, especially in physics, are mathematical theories combined with observations. For instance, Isaac Newton took the mathematical properties of calculus and applied them to our world to create the "idea of gravity", thereby expanding our notion of the law of attraction. Upon closer inspection, the concept of gravity proved to be wrong, since it did not accurately explain Mercury's orbit.[155] The law of gravity was very close to correctly predicting Mercury's orbit, but since it did not account for the relativity of motion, the predicted elliptical path seemed a tiny bit off. Then, Einstein took the mathematical properties of curvature from Bernhard Riemann and applied them to our universe to create the "idea of general relativity", thereby expanding our notion of the law of gravity and more accurately explaining Mercury's orbit. All in all, both gravity and general relativity are completely made up! Useful! But made up.

Through this example, we see theories are "confirmed" once they line up with observations. However, the foundation of these sciences lies in falsifiability; one can seldom prove something to be true. Instead, sciences are an endless cycle of conjectures and refutations, or trial and error. Popper summarizes this well:

> Scientific theories [are] not the digest of observations, but they are inventions—conjectures boldly put forward for trial, to be eliminated if they clashed with observations; with observations which [are] rarely accidental but as a rule undertaken with the definite intention of testing a theory by obtaining, if possible, a decisive refutation.[156]

All in all, conjectures show up as theories, while the refutations are decisive observations that defeat the conjecture. For instance, Newton's theory of gravity was *true*, until its incorrect prediction of Mercury's orbit made it *false*. Einstein's general relativity is currently *true*, until a hypothetical future refutation makes it *false*. As Confucius taught, "It is man who makes truth great, not truth which makes man great."

As stated in the introduction, this endless cycle of conjectures and refutations is well exemplified in elementary statistics. Initially, a hypothesis is proposed. Next, a test is conducted to evaluate the validity of the hypothesis, following the well-known scientific method. Upon concluding the test, you can either (a) fail to reject the hypothesis (i.e., insufficient evidence) or (b) reject the hypothesis. A hypothesis is never explicitly "confirmed." If a hypothesis withstands multiple tests without being rejected, it is then considered "true" or a "natural law."

James Maxwell, the father of electromagnetism, once said: "The true logic of this world is in the calculus of probabilities, [or statistics]."[157] If this proposition holds, then it is simply not possible to prove any scientific theory as "true." As the great physicist and bongo player Richard Feynman once said,

> If we continue to study [physics] more and more, measuring more and more accurately, the [physical] law will continue to become more complicated, not less. In other words, as we study [the] law more and more closely, we find out that it is "falser" and "falser," and the more deeply we study it, and the more accurately we measure, the more complicated the truth becomes. [158]

The only domain where proofs are allowed, or something is said to be true, is mathematics. This is due to the precise nature of the mathematical language. Is one plus one still two? Yes. When the precise language of math is then applied to testable observations, one has physics, or a powerful model used for describing our world. Echoing the words of Feynman once more, "In order to understand physical laws you must understand that they are all some kind of approximation."[159]

Therefore, the "idea of an idea space" may seem "real" with its properties of uncountability and zero measure, but the idea space is but an imagined reality! I can confirm this because I made it up! It is an approximation we can use to make sense of the mind in a way that is congruent with modern physics. In other words, an idea space is new physics. At its heart, it is the trivial solution to Einstein's field equation used to describe the mind, similarly to how Karl Schwarzchild created a solution to describe a stationary star. An idea space is a conjecture waiting for its decisive refutation to prove it wrong.*

The same observational attitude you use in the science of objects, like physics, chemistry, and biology, can also be carried over to the science of the first person, like in an idea space and mindfulness. In mindfulness, we are not meant to take everything at face value. We are tasked with seeing if what others say lines up with our reality.

---

* Personally, I would love for the idea space theory to be proved wrong. How awesome would it be if we found a way to seamlessly transfer thoughts, emotions, sensations, and perceptions to one another?!

For instance, you have probably been told when you close your eyes you see darkness. Take a moment to test this for yourself. Gently close your eyes. Do you see darkness? Or, do you see a fractal landscape filled with different shades of light? When I do this, I see a beautiful light show lighting up my field of vision, like fireworks on the Fourth of July.

Continuing this sentiment, Zen has some powerful teachings, which must be tested by the individual. In a way, Zen is a science—the science of the first person. Zen develops an ineffable approach to understanding how your idea space works. Zen shows us what we are missing when we are not looking, and it should not be dismissed as less of a science than the science of objects simply because an outside observer cannot measure your idea space. Then, as the philosophy of Zen changes and evolves, so too should our understanding of it, like any scientific theory.

Overall, on one hand, we see specific scientific theories, like gravity or buddhahood, are not Lindy-proof, because they do not withstand the test of time. They are human creations bound to change and evolve as time passes. On the other hand, science, or the mechanism of falsification through conjectures and refutations, is entirely Lindy-proof. Science never claims anything to be true and mainly serves as a skeptic's tool to disprove ideas. Thus, science is synonymous with awakening. If awakening is to understand what reality is not, then is that not the exact same nature, or foundation, as science (i.e., falsification)? In life, it is important "to science" instead of "to study science."

The beauty of science, or awakening, is it can be found anywhere. As French poet Guillaume Apollinaire once said, *"L'univers tout entier concentré dans ce vin"* (The whole universe is in a glass of wine) (figure 197).

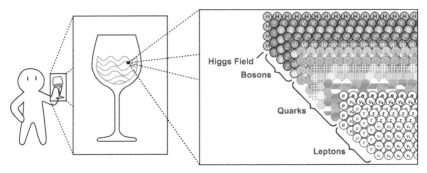

Figure 197. Science is awakening, and it can be found everywhere.

All in all, understanding that words, money, social norms, laws, religions, macropolitical systems, and scientific theories are all imagined realities lifts the next veil of illusion on our Path of Awakening (figure 198). Before, you may have believed that some, if not all, of these concepts were objective truths in our world. Now, these are all fictitious tools that live in the fertile imagination of *sapiens* to make sense of this crazy thing we call life. With this knowledge, we are now ready to awaken.

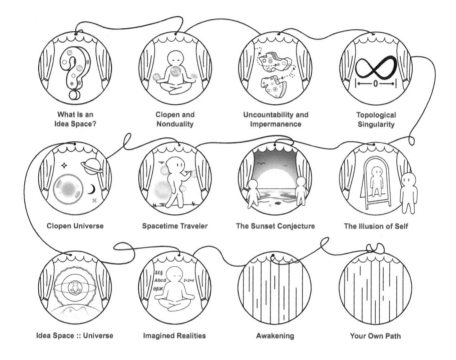

Figure 198. Understanding the difference between objective realities and imagined realities lifts the next veil of illusion on the Path of Awakening.

AWAKENING

Truth is manifest. If put before us naked, it is always recognizable as truth. As Karl Popper said:

> Truth, if it does not reveal itself, has only to be *unveiled*, or dis-covered. Once this is done, there is no need for further

argument. We have been given eyes to see the truth, and the 'natural light' of reason to see it by.[160]

Throughout life, you will come across various truths that will shape the structure of your idea space by building principles. These truths come in different forms, like when a veil of illusion is lifted through a koan. Each koan captures the identity of a particular idea space and transfers the identity from the world of Unknown Unknowns directly into your idea space. When this happens, there is an instant of enlightenment; a koan rips through the fabric of your idea space, like a black hole would spacetime. Thoughts, emotions, sensations, and perceptions of all different kinds are swirled around the koan, as mountains no longer become mountains, trees no longer become trees. You pass through the wormhole of enlightenment as the fabric of your idea space has been completely altered. Afterward, the remnants of this moment will sit in your idea space for an eternity, like a black hole sits in spacetime. A principle has been formed. Once again, mountains become mountains and rivers become rivers. But do not be fooled, these are not the same mountains and rivers as before.

There is nothing "special" to awaken into. You simply awaken to the reality, or present moment; to the same mountains and rivers as before, except they are not the same mountains and rivers as before. Although understanding the differences between objective realities and imagined realities is an important veil of illusion, the whole concept of "Path of Awakening" is an imagined reality in its own right. It's made up. Thus we turn to the final veil of illusion of Act II: There is no awakening for you already are awakened (figure 199).

One does not practice Zen to reach buddhahood, to become buddha, nor to attain *nirvana* (the unconditioned state; the highest peace). As Alan Watts says, "One practices Zen because one is Buddha from the beginning." In other words, one practices Zen because one is, and always has been, awakened.[161] In true Zen and clopen spirit, there is no One Dharma; yet, at the same time, there is One Dharma, the dharma of your reality. We all live in a world based on assumptions. Whether they are true or not is immaterial, as each individual shapes their reality based on the assumptions they choose to embrace.

IMAGINED REALITIES

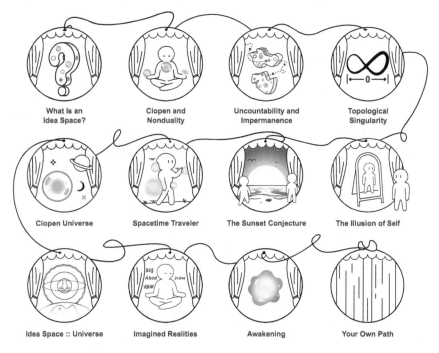

Figure 199. The final veil of illusion of Act II: There is no awakening for you already are awakened.

The beauty of our world is that it is malleable. You have the power to change it simply by breathing, laughing, drinking, sleeping, crying, hearing, seeing, tasting, talking, loving, and doing whatever else makes you human. As Zen master Dzigar Kongtrul Rinpoche said,

> The potential for realization is universal and present for all of us. True benefit will come from your own efforts and realization. For your efforts to bring benefit, you must take your life into your own hands and examine your mind and experience. From this point of view, nobody could be kinder to you than yourself. Nobody could have a greater effect on you or actually do more for you than yourself. . . . If you don't take your life into your own hands, not even the buddhas can make a difference. It's up to you.[162]

Awakening is not an act of doing as much as it is an act of undoing. It is removing our preconceived notions of what we think reality is and instead attuning to what is actually given: the present moment. To achieve awakening, simply realize what reality is not.

It is not your Self. It is not your name. It is not "I." It is not a story about your past. It is not a story about your future. It is not gravity. It is not space. It is not time. It is not the Gregorian Calendar. It is not Summer. It is not December. It is not Friday. It is not seconds. It is not the dollar. It is not English. It is not Bitcoin. It is not laws. It is not the United States. It is not China. It is not the Yen. It is not Republican. It is not Democrat. It is not your title. It is not your company. It is not languages. It is not currencies. It is not Buddha. It is not Christianity. It is not Islam. It is not Judaism. It is not Hinduism. It is not your memories. It is not your expectations. It is not your social status. It is not your cultural identity. It is not your sexual identity. It is not nothing. It is not emptiness. It is not veils of illusion. It is not the Path of Awakening. It is not an idea space.

Even when you take everything you think is real away, there is still an infinite vastness left over, here and now. *That is reality. That is your Non-Self.* Sit in this space and let the warmth consume you. What is left when there is no problem to solve? What is the silence in between sounds? What is your original face before your parents were born?

> *"In the seeing, there is only the seen.*
> *In the hearing, there is only the heard.*
> *In the sensing, there is only the sensed.*
> *In the thinking, there is only the thought."*

End of Act II

Epilogue
THE RETURN OF THE MAGICIAN

"Everyone lives two lives.
The second starts when you realize you only live one."
- Confucius

You close the book and hold it firmly in both hands. The sun is shining as you look out your window at the mountains and rivers—Mount Olympus Mons lies like a backdrop. You contemplate the book and think, *Huh . . . So that was Act II? I didn't even see the final veil of illusion.*

You stand up and take a deep inhalation. As you transition positions, you notice every little sensation present. Your feet feel heavier on the ground. Your spine straightens and lengthens through the crown of your head. Your chest lifts ever so slightly upward. Your belly gets sucked in. You look up and exhale a deep sigh, releasing all the tension in your body—your skeleton holding you strongly.

You start pacing around the room in circles, pondering the only question on your mind: *What was the final trick? What was Act III?* You think back to all the different spots in the book the veil could have been lifted. You place your thumb and index finger on the bridge of your nose as you notice emotions of confusion and curiosity dominating your idea space. However, as soon as you notice them, they disappear into the abyss as sights come to dominate your awareness. You stare at the magician's coin on the desk.

You slowly walk up to it and inspect it once more: "In God We Trust, 2026, the United States of America." You keep flipping the coin back and forth to see if there's anything out of the ordinary with it, but there doesn't seem to be anything fishy. It simply looks like an ancient relic from another time. A time prior to space exploration. A time of borders and countries. A time of union and division. A time very similar to now . . .

You think back to the magic show. The last time you saw the magician, he was standing right next to you. Then—DING—he was gone. *Where did he go? Did he go to the same place as the bunny? How can I find him?* As you ponder the event more and more, you stare back out at the mountains and rivers with Mount Olympus Mons in the backdrop. Subconsciously, you flick the coin into the air—DING.

You put your hand out to catch the coin as you are captured by Mars's beautiful landscape. But . . . the coin never hits your hand. You look into your hand and see nothing. Then, you look up in the air and see the coin still spinning. It's floating there effortlessly (figure 200).

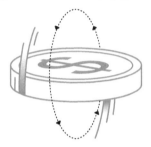

Figure 200. The coin spins.

The coin starts spinning faster and faster. Everything in the background seems to be moving slower and slower. Hover cars move at half speed. Even the birds flap their wings more slowly. You feel your body getting pulled at the top and bottom, while your sides are compressed. Out of the corner of your eye, you see the coin spinning so fast that the space around it seems to be completely distorted. It's as if time and space were being shredded. Then, in one fell swoop, you get spaghettified and sucked into the coin—WOOSH.

You tightly close your eyes. However, instead of darkness, you see a beautiful light show filled with different colors as you pass through a wormhole of time. In an instant, thoughts no longer become thoughts. Emotions no longer become emotions. Sensations no longer become sensations. It's as if, for an instant, time and space were separated. There is no longer time. There is no longer space. There's just this.

Then, in the very edge of your awareness, you hear a familiar sound: water flowing and crashing. You gently unclasp your hands as you feel a soft touch. It's as if you were sitting on a marshmallow cloud. You slowly

EPILOGUE

open your eyes to bright lights that blind you. You put your hands in front of your face.

The world comes rushing in. A majestic river flows before you. The banks are covered by green, lush grass and colorful flowers. Tiny insects fly around. A bunny is sitting right next to you. The bunny tilts its head right and left as it inspects you (figure 201). *Huh—this looks like the same bunny.*

Figure 201. A bunny inspects you.

As you go to pet the bunny, it turns around and hops away. You slowly stand and feel the ever-changing landscape of sensations throughout your body. The bunny turns its head around, as if it's waiting for you. You start to walk forward, noticing every little sensation. *Right foot. Heaviness. Left foot. Lightness. Left foot. Heaviness. Right foot. Lightness.* You start walking toward the bunny as it hops away from you until it stops next to a tree by the river (figure 202). It's a beautiful tree, filled with fractal leaves and branches. The bunny disappears around the tree.

Figure 202. In the distance, the bunny is standing next to the tree by the river.

As you approach, you see a man with a hat sitting beneath the tree. You stare at the man as an impeccable landscape lies before your eyes: a clear river chanting sweet nothings into your ears, the smell of fresh air bringing renewed energy into your nostrils, grand mountains painting an epic backdrop, all while the sun's golden rays seem to be shining directly toward you (figure 203). You stop to take it all in.

Figure 203. The landscape.

"I see you've reacquainted yourself with Mr. Wiggles," says a familiar voice from under the tree.

You've heard this voice before. At first behind the call window. Then from the usher. And finally the magician. *Is this the magician?!* He looks up at you and with a small grin on his face says, "You're probably wondering where we are . . ."

As he stands, he brushes the dirt from his jacket. He takes his hat off and places it onto the ground. The bunny hops into the hat and pops his head back out. The magician stretches his hands in the air, his bones crackling. He lets out a sigh. "Well, congratulations. You're one of the few people who has ever made it to Act III. You're one of the few people who has ever made it into *my idea space*."

The magician starts walking as you follow. "You see," he says, "I've been doing this for a long time." Like a Jedi, he waves his hand in the air, saying, "Plus, times today are completely different compared to when I'm from." The ground starts to shake, and a well appears (figure 204).

EPILOGUE

273

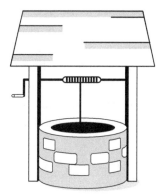

Figure 204. A well appears from the ground.

The magician looks into the well and says, "Back then, we were in the midst of a booming transition. A transition from what was known as the *Industrial Age* to what your history textbooks now call the *Information Age*. Let me show you."

As you look into the well, the magician waves his hand and the water in the well takes on a new form (figure 205). "You see, everyone lives at the center of their own observable universe. Everything you see is in the past. At the center lies your idea space of infinite depth, hidden from the outside world. That's probably why you thought the sun was pointing directly toward you earlier, but in fact, it was actually directed toward me," he says, winking.

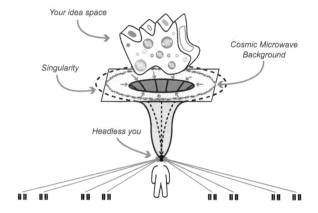

Figure 205. Your idea space lies at the center of your own headless world.

The magician nods his head as he sees the shape your idea space has taken down in the well. He says, "Of course, this is only your idea space. In reality, every spacetime adventurer lives in this same, headless condition. Everyone is at the center of their own observable universe. Everyone experiences their own Singularity Sunset, or 'Source.' From their perspective, at their center lies their own idea space." The magician waves his hand over the well as the water changes shape again (figure 206).

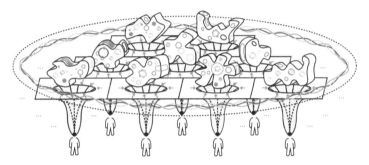

Figure 206. Everyone lives at the center of their own observable universe. At the center lies their own idea space—hidden from the outside world.

The magician continues his explanation, "Although each idea space is hidden from the outside world, we're able to communicate aspects of it through words, songs, equations, body language, etc. When this happens, the real world magic takes place. Personal idea spaces combine to form macro idea spaces. These fantastic beasts still look like nothing to the world, but can dominate the way we live our lives." The magician snaps his fingers, and the water changes once more (figure 207).

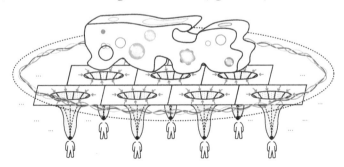

Figure 207. Personal idea spaces combine to form a macro idea space.

EPILOGUE

The magician looks deeply into the well. "Back when I was born," he states, "the dominant macro idea space of the time was the end of the Industrial Age."

He snaps his fingers as the water continuously modifies its form (figure 208). "Of course, macro idea spaces are impermanent, too. With the rise of micro-computing, the internet, and generative artificial intelligence, a new form of currency was able to dominate the communities of my day. Thus, a tear occurred in the macro idea space as the Information Age slowly formed."

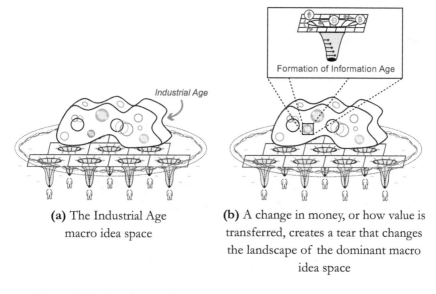

(a) The Industrial Age macro idea space

(b) A change in money, or how value is transferred, creates a tear that changes the landscape of the dominant macro idea space

Figure 208. The Industrial Age turning into the Information Age.

"Afterward," says the magician, "nothing was the same. . . . English no longer was the dominant language. The United States disassembled as the Galactic Trade Federation grew into the dominant intersubjective reality. The new government declared a fresh law of the land, The Rights of Sapiens, that made Arurrak the main form of language and quantum crypto the main currency. It built the reality you've come to know and love on Mars."

The magician pulls you away from the well and claps twice. The well gently sinks back into the ground. He starts walking away as he says underneath his breath, "And so the cycle of empires continues . . ."

You follow, wanting to ask him the only question that's been on your mind: *What's the main trick? What was hiding in plain sight all along?* The magician hears you coming and stops. You almost bump into him.

"Of course," the magician says as he turns around and faces you. "Forgive me. Not many people make it this far, so I tend to forget to reveal the main trick. In all actuality, it's been staring you in the face this whole time. The truth seems to always hide in plain sight—it just looks like nothing. All you have to do is unveil it for its infinite depth to show itself." You stare intensely into the eyes of the magician and gulp: out of fear or excitement, you do not know.

He smiles and then laughs, wiping tears from his eyes, clearly overcome with joy and ecstasy. "It seems so silly to even say, but . . . this book, the concept of an idea space, is the trick!"

The magician extends his right arm. A wand flies out of his sleeve. He points the wand to the sky, and a beautiful medley of clouds appears (figure 209).

Figure 209. Clouds fill the skies.

With a flick of the wand, the magician exalts, "ALOHOMORA!" The clouds start changing shapes, like the water in the well (figure 210). The magician puts his left arm around you while he points at the sky and says, "You see, prior to you even hearing about the idea space, it was an Unknown Unknown in your idea space. It looked like nothing!"

EPILOGUE

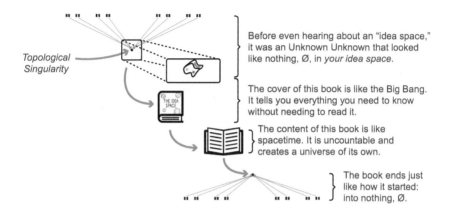

Figure 210. The main trick that was hiding in plain sight all along.

The magician flicks his wand again. "Of course, there was never nothing there! Behind the topological singularity lay something very 'real.' And, as soon as the topological singularity was uncovered, the whole world showed itself to you. The first time you heard about the play, The Idea Space, or saw the cover of the book, was like the Big Bang. It told you everything there was to know without even reading it."

The magician flicks his wand once more. "After my disappearance, you explored the idea space for yourself in Act II. There, the contents were like all of spacetime. They created an uncountable universe inside your mind. You did not take what was said at face value but tested it with your reality to see what principles held true and which didn't."

The magician flicks his wand one final time and yells, "AROMOHOLA!" The clouds gather to a single point of nothingness and disappear, like the bunny from the play—POOF! "And all of it ends just like how it started: into *nothing*."

The ground starts to shake. Even the sky seems to tremble. The magician sighs again. "It is time for you to return to your home. The bridge between idea spaces seldom stays open for this long, so we must be grateful for the opportunity it has given us."

As the world seems to be collapsing on itself (figure 211), you see the magician smile as he sheds a single tear. He shouts over the sound of the fracturing world, "Luckily, Arurrak has a hidden beauty behind it! With koans, I am able to speak to you directly from one point in spacetime to another! The transfer of koans from my idea space to yours will be instantaneous!"

Figure 211. The world, or the magician's idea space, collapses on itself.

His voice drifts away in laughter, "Although . . . what may be a koan for me . . . may not be a koan for you . . . It really depends on the time and place . . ."

You close your eyes and feel yourself spaghettified once more. Thoughts no longer become thoughts. Sensations no longer become sensations. Emotions no longer become emotions. Perceptions are gone. Time and space gone with them—WOOSH!

Suddenly, out of nowhere, you feel the metallic touch of a coin hitting your hand. You are back on Mars. The coin lies in your hand as if nothing has happened. Hover cars move at regular speed. Birds fly as they normally do. Mount Olympus Mons rises in the distance. For some reason, it looks a little different than it did before. It's clearly the same Mount Olympus Mons as before, but you now see it through a strange, new lens (Figure 212)...

Figure 212. Your new view of Mount Olympus Mons.

You glance at your desk and see a note. It reads:

My Dear Volunteer,

Thank you for taking the time to attend the play and read this book. I hope it has helped you as much as it has helped me. Thoughts are hard. The world is hard. Hopefully, now, you're able to see the world yathabhutam—just as it is—without attaching your Self to any thoughts, emotions, sensations, perceptions, or even emptiness.

There is nothing special to awaken into, yet there is something truly delightful to awaken into. Awakening simply means returning to reality, or the present moment. It means not getting lost in thoughts and fantasies about what the world is not. Reality is beautiful. Even when we take all the imagined realities away, there is so much left over. I hope you can now see this for your Non-Self.

As you go forward, remember this: remain clopen for there are uncountably many topological singularities in life. As you come across these veils of illusion, remain vigilant. Spacetime is malleable. Simply by moving you can alter it. Use this to your advantage.

In this vast, boundless world, the power is within you to shape your reality and forge your own path (figure 213). Remember, the world is yours to create and transform, so embrace the beauty of the present and make it happen.

Figure 213. *The final veil of illusion: follow your own path.*

I leave you with one of my favorite koans ... As an Ancient once said: "Everyone lives two lives. The second starts when you realize you only live one." In other words, a transition from your Self to your Non-Self.

With love and care,

The Magician

After reading the note, you pick up the promotional pamphlet that attracted you to the show. A single image and a koan have been added to it (figure 214). Just like that, the magician, himself a mere illusion in your idea space, vanishes into the abyss, never to be seen again.

EPILOGUE

"We do not possess imagination enough
to sense what we are missing."
- Jean Toomer

Figure 214. Your own headless sunset.

FIN

ACKNOWLEDGMENTS

"The meaning of life is life itself."

I tried to include as many names as I could, but I am sure I probably left some off by accident. To everyone whom I have ever met in my life: thank you for making me the person I am today.

In chronological order:

Merci à mes parents, Georges et Sylvie, et mon frère, Romain, pour tout—tout le soutien et l'inspiration depuis le premier jour.

Merci à ma famille élargie, Decrop et Zabner.

Thank you to Mrs. Gillen for teaching me mettā; Mrs. Stachina, Mrs. Lindsay, and Mrs. Wilson, for inspiring my sense of wonder for the sciences; Mrs. Ward and Mrs. Cohen, for making me see the value of arts; and, Mrs. Orens for teaching me how to think.

Thank you to Alec, Charles, and William for your friendship when I did not speak English. Cameron (for the painting!), Colin, Finn, and Rowan, for an international friendship. Thank you to Asim, Bobby, David, Marc, Neeraj, and Steven for the endless poker games.

To the Marsh family, Alex, Bobby, Eric, Matt, and Michele, for a second home. To the Rodale family, Anthony, Coco, Florence, and Marlow, for all the holidays.

For the countless days of playing Super Smash Bros Melee and feasting: Arun, Connor, Harvey, Jon, Karthik, Neil, Ryan, and Thomas.

To the various great high school communities. E-crew: Adam, Jake, Kristyan, Matt, and Nick. LMT: Dalton, Gabe, Justin, Luke, Mike, and Seth. And to the other amazing high school friends: Darren (I still owe you gum), Devon, Greg, Kaitlyn, Mike, and Mike Borgioni (RIP).

Thank you for your friendship during the college years: Adrea, Brad, Chance, Claire, Danny, Greg, Jess, Josh, Kevin, Lucas, Luke, Marco, Pete, Rich, Samii, Shannon, and Tim. And for the time post-graduate: Joe, Kyle, and Ryan.

Pablo, for the guidance in Spain. Farah, for being a soulmate across the seas.

For the great memories in DC: Alli, Allie, Caleb, Katie B., Kane, Katie G., Kevin, Megan, Michael, Molly, Sara, and Winters. Ball is life Umair.

To the ski crew: Alec, Anders, Chris, CJ, Jack, Jake, and Joey.

Jake, Jeremy, Martin, Tommy, Spencer, and Zach, for being the best co-inventors I could ask for. To Annaliza, Earl, and Joe for the friendship and mentorship. Madeline, Michael, Nikhil, and Veronique, for the wellness crew.

Emanuel, for inspiring me to write a book. Shane, for being the first person to listen to me speak about my book. Ryan, for sharing a love of travel, writing, and Zen, and for reading early versions of my book.

Matt, RJ, and Ryan, for the continuous friendship and support, and for listening to a four-day lecture on idea space.

To Soris (Ashley), Strahd (Cam), Topher (Colin), and Xander (Sanda) for an adventure unlike any other.

To Alex, Kyle, Shawn, and Tucker for the frolf. Xavier, for the gym and ball is life.

Thank you to David for being a guiding light in all things physics and for listening to me blabber about nonsense for countless hours. Nikki, for allowing me to speak in front of an audience when no one else would. Tim, for engaging in spirited conversations on physics with a total stranger.

Dani, for creating the bubbles. Tea, for your magic. Onur, for literally everything. Basak, for the awesome website design; and Mahtab, for bringing the website to life.

Lynette and Chip for the killer photos!

Thank you to Mary for being an amazing coach and helping frame the book. Thank you to Hal for providing thoughtful, meticulous feedback and for taking a chance on me—you're a class-act editor. ChatGPT for bouncing around ideas in the editing stage. Thank you to Shannon for taking the time to copyedit the manuscript and make sure I dotted all my I's and crossed all my T's.

Kevin and team, for creating an amazing animation; David for scoring it perfectly; and Simon, for voicing the magician.

Thank you, Sil, for creating amazing social media content; and, Eunice for the constant support. Liv for your willingness to take on random one off projects. Mathew for taking on a managerial role.

Lastly, I extend my heartfelt gratitude to those I never had the chance to meet, yet who have served as an endless source of inspiration for this book, especially Albert Einstein, Elon Musk, Joe Rogan, Richard Feynman, Steve Jobs, and the countless others listed in the bibliography.

LEXICON

"The task is not to see what has never been seen before, but to think what has never been thought before about what you see every day."
- Erwin Schrodinger

Accretion disk: a disk of gas that surrounds a stellar object, like a protostar or black hole. Friction in the disk makes the gas gradually spiral inward and *accrete* onto the hole or star.
Anattā: selflessness.
Anicca: impermanence.
Antimatter: the "anathema" to ordinary matter. Every particle has its antiparticle counterpart (e.g., electron vs. anti-electron). The main difference is that it has opposite charge.
Arahant: fully enlightened individual.
Atom: basic building block of matter. Each atom consists of a nucleus with a positive charge and a surrounding cloud of electrons with negative charge.
Awakening: realizing what the world is not; returning to reality, or the present moment.
Baryons: a combination of three quarks.
Beginner's mind: the empty idea space. A great place to start a new endeavor.
Bhikkhu: someone on the Path of Awakening or enlightenment.
Big Bang: the start of the universe. The edge of our observable universe.
Big Bang model: the first 380,000 years of the universe. Includes events from the Big Bang to Recombination.
Bijection: a function, or map, that is both one-to-one and onto.
Black hole: An object (caused from the implosion of a star) into which matter and energy can fall, but nothing can ever escape. A rip through the fabric of spacetime.
Bosons: force-carrying particles.

Brainblast: the equivalent of a neutron star in an idea space that occurs after a moment of enlightenment.
Brown dwarf: a failed star that never hit main sequence, or has a very short main sequence.
Buddha: awakened one. A man or woman of no rank. Buddha is already within us.
Cantor Set: a mathematical object that is uncountable and has zero measure.
Cantor's diagonal argument: no matter how many real numbers you list out, even an infinite amount, I will always be able to choose one you did not list.
Cardinality: the number of elements in a set. Two sets have equal cardinality if and only if there is a bijection between both sets.
Choice point: an event where a decision needs to be made—happens at every moment.
Clopen: a space that is open and closed at the same time.
Closed idea space: an idea space that contains all its limits, or thoughts. Focused awareness.
Closed society: a society which believes that the ruling intersubjective reality of the time is more important than the individuals of that intersubjective reality. Otherwise known as the magical, tribal, or collectivist society.
Complement: what remains of a set when one takes part of the original set out. Two sets are complements of one another if they make up the whole set once combined.
Consciousness: light one shines onto an idea space. An enigmatic mystery.
Cosmic Microwave Background (CMB): the outermost veil of our observable universe. Light behind this region is prevented from escaping to the rest of our universe.
Countable: there exists a bijection between the natural numbers and whatever set you are counting.
Dark energy: a mysterious form of energy that causes the constant expansion of space. It is ubiquitous through every point in space and time. It is what makes our universe impermanent. Dark energy makes up 68 percent of our universe.
Dark matter: mysterious matter that makes up 27 percent of observable universe and is responsible for gravitational seeds that form galaxies and keep them together.

Devil's Staircase: an object that has global curvature but remains flat throughout.
Dharma: the law; the truth; the teachings of the Buddha.
Dukkha: suffering; frustration; unsatisfactoriness; stress.
Einstein Equivalence Principle: in small-enough regions of spacetime, the laws of physics reduce to those of special relativity; it is impossible to detect the existence of a gravitational field by means of local experiments.
Electric charge: the property of a particle or matter by which it produces and feels electric forces.
Electron: a fundamental particle with negative charge that populates outer regions of atoms.
Elementary ideas: thoughts, emotions, sensations, and perceptions.
Empty set, ∅: nothing or emptiness. Looks like zero, 0, but it is not zero! The empty set is clopen.
Enlightenment: analogous to a supernova. An aha moment caused by the creation of brainblast or koan. The moment lasts an instant, but the remnants last an eternity. One of the keys to awakening.
Euclidean geometry: geometry of perfect lines, squares, and circles. Used in a flat spacetime.
Event: a point in spacetime.
Everything: the complement to nothing, or the empty set. "Everything," as a grouping, is clopen.
Fermions: matter particles (e.g., quarks and leptons).
Fission: the breaking up of larger atomic nuclei into several smaller ones.
Five Aggregates: thoughts, emotions, sensations, perceptions, and consciousness
Fractals: the study of roughness. Usually associated with self-similar patterns at different scales.
Free will: Do we have control of our actions? The situation is clopen: we do and we don't.
Function: a map or transformation that takes a point from set A to another point in set B.
Fundamental ideas: Unknown Unknowns, Known Knowns, Known Unknowns, Unknown Knowns.
Fusion: the merger of two atomic nuclei into one larger one.
General Relativity: Einstein's baby. The warpage of spacetime by matter or energy to create the illusion of gravity.

Gluon: force-carrying particle associated with the strong force.

Grand Unification Theory: a time in the early universe where the strong force, the electromagnetic force, and the weak force were all one. Furthermore, quarks and leptons were one and the same.

Gravitational wave: a ripple of spacetime curvature that travels at the speed of light.

Graviton: hypothetical particle associated with gravitational waves.

Hadrons: any combination of quarks.

Headless way: mindfulness technique developed by Douglas Harding to see your Non-Self and answer the koan: *What is my original face, before my parents were born?*

Higgs boson: the particle that gives mass to matter (minus neutrinos) and $W^{\pm}$ and Z^0 bosons.

Horizon: the surface of a black hole; the point of no return.

Hubble Radius: the edge of our universe *now*.

Idea space :: universe: your idea space is a reflection of your observable universe.

Idea space: a place that consists of thoughts, emotions, sensations, perceptions, and the empty set. It is unique to you, uncountable, and has zero measure. Your idea space lives at the center of your own observable universe.

Identity: the pure essence of a particular idea. See koan.

Illusion of Self: "I," your name or your identity, is simply an amalgamation of your thoughts, emotions, sensations, and perceptions. It is an idealization others use to approximate your Non-Self.

Impermanence: constant state of flux. The universe and our idea space are impermanent. Synonymous with uncountable.

Inflation: period in the early universe, where the universe super-cooled and expanded at least 10^{26} fold.

Information Age: the era after the Industrial Revolution. Made possible through the advancements in micro-processing.

Injection: a function, or map, that is one-to-one.

Insights: the equivalent of a white dwarf in an idea space.

Intersubjective realities: a type of macro idea space that exists within the interconnected communication systems of humans. Examples include words, money, governments, etc. An intersubjective reality does not disappear if one person stops believing in it.

Known Knowns: ideas you know to be true relative to the world.

Known Unknowns: ideas that you know, but the world doesn't.
Koan: a phrase, story, dialogue, question, statement, picture, feeling, or sensation that perfectly captures the identity of an idea space, or a moment in space and time. It is a tool that is used to achieve sudden, intuitive enlightenment. It creates principles in your idea space that can be used to traverse through your idea space, like a wormhole.
Light cone: defines how light comes to and leaves a particular event. The edge of your past light cone is your observable universe.
Lindy Test: an exercise to see if an idea is nonperishable and withstands the test of time. The longer the idea persists, the more likely it will remain for the foreseeable future.
Macro idea space: a mixing of personal idea spaces between people. This can be a verbal or nonverbal interaction.
Main sequence: when a star initiates fusion.
Meditation: the formal practice of mindfulness.
Meson: a combination of a quark/anti-quark pair.
Mettā: loving-kindness.
Mindfulness: awareness that arises through paying attention, on purpose, in the present moment, nonjudgmentally. Awareness of our idea space in between topological singularities.
Nāmarūpa: mind-body duality. A clopen phenomena.
Nebula: a cloud of gas that creates stars and galaxies.
Nen: thought impulse.
Neutrinos: the most exotic fermion. Its mass is dubiously small, and it does not interact with the Higgs Field.
Neutron: a particle that forms the nucleus of an atom along with a proton. A neutron has zero charge and is made up of two down quarks and one up quark.
Neutron star: the death of a large star.
Nirvana: unconditioned state; the highest peace.
Nonduality: synonymous to clopen. Whenever we group, we get a duality. Nonduality involves taking a step back and seeing the duality live together harmoniously.
Non-Euclidean geometry: geometry of fractals and curved surfaces. Used to describe general relativity.
Non-Self: realizing "I," your name, is simply an appearance in your idea space. It is one layer of your true Self, which in all actuality consists of all

the layers of your observable universe, as everyone experiences their own Sunset Singularity. Your Non-Self is who you are at zero measure—prior to any thoughts, emotions, sensations, or perceptions.

Nucleus: core of an atom, which consists of neutrons and protons.

Observable universe: a relatively small, finite portion of the grander universe. It is the edge of our past light cones. Although the universe has no center, everyone is at the center of their own observable universe. Everything we see is in the past.

One-Nen: staying with the first thought impulse, or staying in phenomena.

Open idea space: an idea space that does not contain all its limits. Choiceless Awareness.

Open society: a society that believes it is the purpose of the state, or intersubjective reality, to protect its individuals. In this society, individuals are confronted with personal decisions and there is no closure to the outside world. In an open society, members strive to rise socially.

Particle-wave duality: the ability for any particle to act like a wave and particle at the same time.

Personal Idea Space: your own idea space.

Photon: the particle associated with light.

Practice: the embodiment of an approach to a concept.

Principle: caused by the formation of a koan in your idea space. A fundamental truth or proposition that serves as the foundation for a system of belief or behavior or for a chain of reasoning.

Quantum mechanics: the laws of physics that govern the small. Among the phenomena are particle wave-duality, the uncertainty principle, and Hawking radiation.

Quarks: the building blocks of matter that make up particles like protons and neutrons.

Recombination: the point where electrons and photons decoupled. Thus, electrons attached to the nucleus, and atoms were formed. With this, light was able to travel the universe freely, and so the universe became transparent.

Reference frame: three spatial coordinates *(x, y, z)* combined with a temporal coordinate *(t)*.

Samādhi: pure concentration; flow. This can be achieved by seeing the nondual nature of reality.

Schwarzschild radius: the maximum radius of a star, based on its mass, at which it implodes into a black hole.

Science: a cycle of conjectures and refutations set on the principle of falsification used to more clearly see what the world is not. Synonymous with awakening.

Science of objects: approximations used for observations that can be measured. Includes the fields of physics, chemistry, biology, and the other natural sciences.

Science of the first person: approximations used for observations that have zero measure, like your idea space.

Set: a group of elements

Singularity: place in spacetime where curvature is so strong that the laws of general relativity break down. As Stephen Hawking states, "Singularities are outside the presently known laws of physics."

Singularity Sunset: like a sunset, the edge of our own observable universe, the Big Bang, is unique to us. In other words, the gravitational effects of the Big Bang, the birth of our universe and the beginning of time, impacts everyone uniquely.

Source: another name for the Big Bang at the edge of your observable universe.

Spacetime: the four-dimensional fabric of space and time. Space and time are one and the same.

Special relativity: how one calculates distance, in a flat spacetime. Spacetime is locally flat with global curvature; the rules of special relativity always apply: when one person moves, they contract in the direction of motion relative to a stationary individual. Furthermore, time contracts for the moving individual compared to the stationary one.

Standard model of physics: paradigm for dealing with all the subatomic particles in the universe.

Stoicism: mindfulness technique used to limit negative emotions.

Sunset Conjecture: (a) everyone lives at the center of their own observable universe with their own Singularity Sunset, (b) at the center lies your idea space of uncountable depth and zero measure.

Supernova: a gigantic explosion of a massive dying star.

Surjection: a function, or map, that is onto (i.e., all of A onto all of B).

Theory of everything: hypothetical time in the super early universe where all four forces were one.

Topological Singularity: a set of objects that are uncountable, have zero measure, and look like nothing, ∅ (e.g., the Cantor Set, an idea space, etc.).

Uncertainty principle: the more precisely one measures the particular location of a particle, the more difficult it will be to measure its momentum; and vice versa.

Uncountable: not countable. There is no bijection between a set and the natural numbers. Spacetime and your idea space are uncountable. Synonymous with impermanent.

Universe: everything in our observable universe and then some.

Unknown Knowns: ideas that the world knows but that seem like variables to you. When you "kinda" know something.

Unknown Unknowns: ideas that no one in the world knows (e.g., gravity before gravity, idea space before idea space, etc.).

Veils of Illusion: when an Unknown Unknown enters an unsuspecting idea space. It often takes the form of a topological singularity (it looks like nothing on the surface; however, as soon as you uncover it, it demonstrates uncountable depth).

Weak equivalence principle: the motion of freely falling particles is the same in a gravitational field and a uniformly accelerated frame, in small-enough regions of spacetime.

White dwarf: the death of a small star.

$W\pm$ and Z^0 Boson: force-carrying particle associated with the weak force.

Worldline: the path a particle or human takes through spacetime.

Wormhole: a "handle" in topology that connects to disparate points in the universe.

Yathabhutam: just as it is.

Zero measure: an object so small, because it is of size zero, 0, that it looks like nothing. Remember, $0 \neq \emptyset$!

BIBLIOGRAPHY

"A reader lives a thousand lives before he dies.
The man who never reads lives only once."
- Game Of Thrones

Please find below the books, papers, and articles that consciously and subconsciously inspired *The Idea Space*. **Sources in bold are classics**. Read the classics; they're classics for a reason.

MINDFULNESS

Brewer, Judson et al. "Meditation experience is associated with differences in default mode network activity and connectivity." *PNAS* 50, no. 50 (2011): 20, 254–260, 259.

Csikszentmihalyi, Mihaly. "Flow: The Psychology of Optimal Experience," (Harper & Row, 1990).

"Jon Kabat-Zinn: Defining Mindfulness." *Mindful*, accessed December 1, 2021. https://www.mindful.org/jon-kabat-zinn-defining-mindfulness/.

Goldstein, Joseph. *Mindfulness: A Practical Guide to Awakening* (Sounds True, 2016).

Goldstein, Joseph and Jack Kornfield. *Seeking the Heart of Wisdom: The Path of Insight Meditation* (Shambhala Classics, 1987).

Lutang, Lin. *The Importance of Living* (Harper, 1937).

Harris, Sam. *Waking Up: A Guide to Spirituality Without Religion* (Simon and Schuster, 2014).

Harris, Sam. *Waking Up*. Waking Up LLC, Version 2.2. *https://www.wakingup.com/*, https://apps.apple.com/us/app/waking-up-guided-meditation/id1307736395.

Harding, Douglas. *On Having No Head: Zen and the Rediscovery of the Obvious* (The Buddhist Society, 1961).

Huikai, Wumen. *The Gateless Gate* (Boomer Books, 2008).

Mayer, John. "What Is Emotional Intelligence?" UNH Personality Lab. 8, 2004. https://scholars.unh.edu/personality_lab/8.
Pirsig, Robert. *Zen and the Art of Motorcycle Maintenance* (Morrow Modern Classics, 1974).
Rubin, Rick. *The Creative Act: A Way of Being* (Penguin Press, 2023).
Sekida, Katsuki. *Zen Training* (Shambhala Classics, 1985).
Tou, Hseuh. *The Blue Cliff Record*. Translated by Thomas Cleary & J.C. Cleary (Shambhala, 2005).
Watts, Alan. *The Way of Zen* (Vintage Books. 1957).

PSYCHOLOGY

Kahneman, Daniel. *Thinking, Fast and Slow* (Farrar, Straus, and Giroux, 2011).
Frankl, Viktor. *Man's Search for Meaning* (Beacon Press, 1959).
Sapolsky, Robert. *Behave: The Biology of Humans at Our Best and Worst* (Penguin Random House, 2017).

PHILOSOPHY

Jorgenson, Eric. *The Almanack of Naval Ravikant* (Magrathea Publishing, 2020).
Popper, Karl. *Conjectures and Refutations* (Routledge & Kegan Paul, 1963).
Popper, Karl. *The Open Society and Its Enemies* (Princeton University Press, 1945).
Taleb, Nassim. *The Black Swan: The Impact of the Highly Improbable* (Second edition, Random House, 2007).
Taleb, Nassim. *Skin in the Game: Hidden Asymmetries in Daily Life* (Random House, 2018).

POETRY

Whyte, David. *Consolations: The Solace, Nourishment and Underlying Meaning of Everyday Words* (Many Rivers Press, 2019).

MATHEMATICS

Pugh, Charles. *Real Mathematical Analysis* (Springer, 2002).
Falconer, K.J. *The Geometry of Fractal Sets* (Cambridge University Press, 1985).
Kosniowki, Czes. *A First Course in Algebraic Topology* (Cambridge University Press, 1980).
Mandelbrot, Benoit. *The Fractal Geometry of Nature* (Freeman, 1977).
Dummit, David and Richard Foote. *Abstract Algebra*, Third edition (John Wiley & Sons, 2004).

GENERAL SCIENCE

Darwin, Charles. *The Origin of Species* (1859).
Dawkins, Richard. *The Selfish Gene* (Oxford University Press, 1976).
Greenfield, Ben. *Boundless: Upgrade Your Brain, Optimize Your Body, and Defy Aging* (Victory Belt Publishing, 2020).
Plomin, Robert. *Blueprint: How DNA Makes Us Who We Are* (MIT Press, 2018).
Nestor, James. *Breath: The New Science of a Lost Art* (Riverhead Books, 2020).
Sinclair, David. *Lifespan: Why We Age—and Why We Don't Have To* (Atria Books, 2019).
Strassman, Rick. *DMT: The Spirit Molecule* (Park Street Press, 2001).

HISTORY

Davidson, James, and William Rees-Mogg. *The Sovereign Individual: Mastering the Transition to the Information Age* (Simon and Schuster, 1999).
Diamond, Jared. *Guns, Germs, and Steel: The Fates of Human Societies* (Norton, 1997).
Harari, Yuval. *Sapiens: A Brief History of Humankind* (HarperCollins, 2015).
Pickover, Clifford. *The Math Book: From Pythagoras to the 57th Dimension, 250 Milestones in the History of Mathematics* (Sterling, 2009).
Plato. *Republic*. Translated by Benjamin Jowett (Independently published, 2020).

Ridley, Matt. *The Rational Optimist: How Prosperity Evolves* (HarperCollins, 2010).

Walter Isaacson, *Einstein: His Life and Universe* (New York: Simon & Schuster, 2007).

PHYSICS

Borde, Arvind, Alan Guth, and Alexander Vilenkin. "*Inflationary Spacetimes Are Not Past-Complete.*" Physical Review Letters 90 (2003).

Carroll, Sean. The Big Picture: On the Origins of Life, Meaning, and the Universe Itself (Dutton, 2016).

Carroll, Sean. *From Eternity to Here: The Quest for the Ultimate Theory of Time* (Dutton, 2010).

Carroll, Sean. *Something Deeply Hidden: Quantum Worlds and the Emergence of Spacetime* (Dutton, 2019).

Carroll, Sean. *Spacetime and Geometry: An Introduction to General Relativity* (Cambridge University Press, 2019).

Chuss, David. "*Measuring Cosmic History: Tools for Understanding the Universe.*" IBHA Conference, 2018.

Collins, Christopher, and Stephen Hawking. "*Why Is the Universe Isotropic?*" Astrophysical Journal, 180 (1973): 317–34.

"DOE Explains . . . Neutrinos." *Office of Science*, accessed June 11, 2022. *https://www.energy.gov/science/doe-explainsneutrinos*.

Einstein, Albert. *Relativity: The Special & General Theory* (Henry Holt and Company, 1920).

Feynman, Richard. The Feynman Lectures on Physics Volumes I–III (Basic Books, 1965).

Griffiths, David, and Darrell Schroeter. *An Introduction to Quantum Mechanics*, Third Edition (Cambridge University Press, 2018).

Hawking, Stephen. The Illustrated a Brief History of Time (Bantam, 1988).

Hawking, Stephen, and George Ellis. The Large Scale Structure of Space-Time (Cambridge University Press, 1973).

"How Old Is the Milky Way?" *European Southern Observatory*, accessed June 11, 2022.
https://www.eso.org/public/usa/news/eso0425/.

"Hubble Reveals Observable Universe Contains 10 Times More Galaxies

Than Previously Thought." NASA, accessed June 11, 2022. https://www.nasa.gov/feature/goddard/2016/hubble-reveals-observable-universe-contains-10-times-more-galaxies-than-previously-thought.

Kolb, Edward, and Michael Turner. *The Early Universe* (Westview Press, 1990).

Lincoln, Don. *"If the Universe Is Only 14 Billion Years Old, How Can It Be 92 Billion Light Years Wide?"* Fermilab, accessed June 11, 2022. https://www.youtube.com/watch?v=vIJTwYOZrGU.

Mack, Katie. *The End of Everything (Astrophysically Speaking)* (Scribner, 2020).

Madau, Piero, and Mark Dickinson. *"Cosmic Star Formation History."* Annual Review of Astronomy and Astrophysics (ARAA) 52 (2014).

Misner, Charles, Kip Thorne, and John Wheeler. *Gravitation* (W. H. Freeman and Company, 1973).

Mo, Houjun, Frank van den Bosch, and Simon White. *Galaxy Formation and Evolution.* (Cambridge University Press, 2010).

Penrose, Roger. *"Gravitational Collapse and Spacetime Singularities."* Physics Review Letters 14, 57 (1965).

Peskin, Michael, and Daniel Shroeder. *An Introduction to Quantum Field Theory* (CRC Press, 1995).

"Planck Cosmic Recipe." *European Space Agency*, accessed June 11, 2022. https://www.esa.int/ESA_Multimedia/Images/2013/03/Planck_cosmic_recipe.

Ryan, Sean, and Andrew Norton. *Stellar Evolution and Nucleosynthesis* (Cambridge University Press, 2010).

Schröder, K.P., and Robert Smith. "Distant Future of the Sun and Earth Revisited." *Monthly Notices of the Royal Astronomical Society* 386, no. 1 (2008): 155–63.

"The Milky Way." NASA, accessed June 11, 2022. https://imagine.gsfc.nasa.gov/features/cosmic/milkyway_info.html.

Thorne, Kip. *Black Holes and Time Warps: Einstein's Outrageous Legacy* (Norton, 1994).

Weinberg, Steven. *Gravitation and Cosmology: Principles and Applications of the General Theory of Relativity* (John Wiley & Sons, 1972).

OTHERS

Nakatomo, Satoshi. "Bitcoin: A Peer-to-Peer Electronic Cash System." 2008. www.Bitcoin.org.

Szabo, Nick. "Shelling Out: The Origins of Money." *Satoshi Nakamoto Institute*, accessed June 11, 2022. https://nakamotoinstitute.org/shelling-out/.

"The World Factbook: United Kingdom." *CIA*, accessed June 11, 2022. https://www.cia.gov/the-world-factbook/countries/united-kingdom/.

NOTES

"Good artists copy.
Great artists steal."
- Picasso

[1] Tou, Hseuh. *The Blue Cliff Record*. Translated by Thomas Cleary & J.C. Cleary (Shambhala, 2005), 179.
[2] Ibid., xiv.
[3] Harari, Yuval. *Sapiens: A Brief History of Humankind* (HarperCollins, 2015), 117.
[4] Watts, Alan. *The Way of Zen* (Vintage Books, 1957), 171.
[5] Kahneman, Daniel. *Thinking, Fast and Slow* (Farrar, Straus, and Giroux, 2011), 44.
[6] Tou, Hseuh. *The Blue Cliff Record*. Translated by Thomas Cleary & J.C. Cleary (Shambhala, 2005), 429.
[7] Watts, Alan. *The Way of Zen* (Vintage Books, 1957), 51.
[8] Dawkins, Richard. *The Selfish Gene* (Oxford University Press, 1976), 246.
[9] Huikai, Wumen. *The Gateless Gate* (Boomer Books, 2008), 1.
[10] Goldstein, Joseph. *Mindfulness: A Practical Guide to Awakening* (Sounds True, 2016), 219.
[11] Watts, Alan. *The Way of Zen* (Vintage Books, 1957), 73.
[12] Harris, Sam. *Waking Up: A Guide to Spirituality Without Religion* (Simon and Schuster, 2014), 57.
[13] Huikai, Wumen. *The Gateless Gate* (Boomer Books, 2008), 59.
[14] Ibid., 65.
[15] Kosniowki, Czes. *A First Course in Algebraic Topology* (Cambridge University Press, 1980), 11.
[16] Popper, Karl. *Conjectures and Refutations* (Routledge & Kegan Paul, 1963), 38.
[17] Rubin, Rick. *The Creative Act: A Way of Being* (Penguin Press, 2023), 43.
[18] Harris, Sam. *Waking Up: A Guide to Spirituality Without Religion* (Simon and Schuster, 2014), 37.
[19] Goldstein, Joseph. Mindfulness: A Practical Guide to Awakening (Sounds True, 2016), 30.
[20] Ibid., 273.
[21] Ibid., 158.

[22] Nishimoto, Shinji and Yu Tagi. *High-resolution image reconstruction with latent diffusion models from human brain activity*, bioRxiv, 10.1101 (2022).

[23] Goldstein, Joseph. *Mindfulness: A Practical Guide to Awakening* (Sounds True, 2016), 108.

[24] Ibid., 75.

[25] Tou, Hseuh. *The Blue Cliff Record*. Translated by Thomas Cleary & J.C. Cleary (Shambhala, 2005), 474.

[26] Lutang, Lin. *The Importance of Living* (Harper, 1937), 5.

[27] Watts, Alan. *The Way of Zen* (Vintage Books, 1957), 144.

[28] Tou, Hseuh. *The Blue Cliff Record*. Translated by Thomas Cleary & J.C. Cleary (Shambhala, 2005), 236.

[29] Ibid., 81.

[30] Huikai, Wumen. *The Gateless Gate* (Boomer Books, 2008), 41.

[31] Richard P. Feynman, Robert B. Leighton, and Matthew Sands, *Characteristics of Force,* The Feynman Lectures on Physics, Volume 1, Online edition, Caltech, accessed May 17, 2021. https://www.feynmanlectures.caltech.edu/I_12.html.

[32] Goldstein, Joseph and Jack Kornfield. *Seeking the Heart of Wisdom: The Path of Insight Meditation* (Shambhala Classics, 1987), 151.

[33] Goldstein, Joseph. *Mindfulness: A Practical Guide to Awakening* (Sounds True, 2016), 97.

[34] Mandelbrot, Benoit. *The Fractal Geometry of Nature* (Freeman, 1977), 15.

[35] "The World Factbook: United Kingdom." *CIA*, accessed June 11, 2022. https://www.cia.gov/the-world-factbook/countries/united-kingdom/.

[36] Pirsig, Robert. *Zen and the Art of Motorcycle Maintenance* (Morrow Modern Classics, 1974) 57.

[37] Harris, Sam. *Waking Up: A Guide to Spirituality Without Religion* (Simon and Schuster, 2014), 142.

[38] Ibid., 146.

[39] Harding, Douglas. *On Having No Head: Zen and the Rediscovery of the Obvious* (The Buddhist Society, 1961), 11.

[40] Ibid., 2-3.

[41] Mack, Katie. *The End of Everything* (Astrophysically Speaking) (Scribner, 2020), 17.

[42] Carroll, Sean. *From Eternity to Here: The Quest for the Ultimate Theory of Time* (Dutton, 2010), 78.

[43] Ibid., 95.

[44] Mack, Katie. *The End of Everything* (Astrophysically Speaking) (Scribner, 2020), 20.

[45] Rubin, Rick. The Creative Act: A Way of Being (Penguin Press, 2023), 14.

[46] Hawking, Stephen, and George Ellis. *The Large Scale Structure of Space-Time* (Cambridge University Press, 1973), 3.

[47] "COBE (Cosmic Background Explorer)." COBE (Cosmic Background Explorer). NASA Goddard Space Flight Center. Accessed May 17, 2023. https://lambda.gsfc.nasa.gov/product/cobe/.

[48] Lincoln, Don. "If the Universe Is Only 14 Billion Years Old, How Can It Be 92 Billion Light Years Wide?" *Fermilab*, accessed June 11, 2022. https://www.youtube.com/watch?v=vIJTwYOZrGU.

[49] Carroll, Sean. *From Eternity to Here: The Quest for the Ultimate Theory of Time* (Dutton, 2010), 58.

[50] Mack, Katie. *The End of Everything* (Astrophysically Speaking) (Scribner, 2020), 97.

[51] Carroll, Sean. *From Eternity to Here: The Quest for the Ultimate Theory of Time* (Dutton, 2010), 60.

[52] Thorne, Kip. *Black Holes and Time Warps: Einstein's Outrageous Legacy* (Norton, 1994), 337.

[53] Mack, Katie. *The End of Everything* (Astrophysically Speaking) (Scribner, 2020), 83.

[54] Lincoln, Don. "If the Universe Is Only 14 Billion Years Old, How Can It Be 92 Billion Light Years Wide?" *Fermilab*, accessed June 11, 2022. https://www.youtube.com/watch?v=vIJTwYOZrGU.

[55] Kolb, Edward, and Michael Turner. *The Early Universe* (Westview Press, 1990), 76.

[56] Popper, Karl. *Conjectures and Refutations* (Routledge & Kegan Paul, 1963), 242.

[57] Weinberg, Steven. *Gravitation and Cosmology: Principles and Applications of the General Theory of Relativity* (John Wiley & Sons, 1972), 3.

[58] Walter Isaacson, Einstein: His Life and Universe (New York: Simon & Schuster, 2007), 193.

[59] Thorne, Kip. *Black Holes and Time Warps: Einstein's Outrageous Legacy* (Norton, 1994), 31.
[60] Ibid., 551.
[61] Carroll, Sean. *From Eternity to Here: The Quest for the Ultimate Theory of Time* (Dutton, 2010), 107.
[62] Thorne, Kip. Black Holes and Time Warps: Einstein's Outrageous Legacy (Norton, 1994), 127.
[63] Ibid., 111.
[64] Carroll, Sean. *Spacetime and Geometry: An Introduction to General Relativity* (Cambridge University Press, 2019), 50.
[65] Ibid., 49-50.
[66] Richard P. Feynman, Robert B. Leighton, and Matthew Sands, *Characteristics of Force*, The Feynman Lectures on Physics, Volume 1, Online edition, Caltech, accessed May 17, 2021. https://www.feynmanlectures.caltech.edu/I_12.html.
[67] Richard P. Feynman, Robert B. Leighton, and Matthew Sands, *Relativistic Energy and Momentum*, The Feynman Lectures on Physics, Volume 1, Online edition, Caltech, accessed May 17, 2021. https://www.feynmanlectures.caltech.edu/I_12.html.
[68] Walter Isaacson, *Einstein: His Life and Universe* (New York: Simon & Schuster, 2007), 130.
[69] "How long does a bacterium live?," Science Focus, accessed October 14, 2022, https://www.sciencefocus.com/nature/how-long-does-a-bacterium-live/.
[70] Sekida, Katsuki. Zen Training (Shambhala Classics, 1985), 111.
[71] Goldstein, Joseph. *Mindfulness: A Practical Guide to Awakening* (Sounds True, 2016), 193.
[72] Watts, Alan. *The Way of Zen* (Vintage Books, 1957), 53.
[73] Hawking, Stephen. *The Illustrated a Brief History of Time* (Bantam, 1988), 229.
[74] "DOE Explains . . . Neutrinos." *Office of Science*, Accessed June 11, 2022. *https://www.energy.gov/science/doe-explainsneutrinos*.
[75] "DOE Explains...The Standard Model of Particle Physics," U.S. Department of Energy, accessed June 11, 2022. https://www.energy.gov/science/doe-explainsthe-standard-model-particle-physics.

[76] "Planck: Cosmic Recipe," European Space Agency (ESA). Accessed July 11, 2022. https://www.esa.int/ESA_Multimedia/Images/2013/03/Planck_cosmic_recipe.

[77] Plomin, Robert. Blueprint: *How DNA Makes Us Who We Are* (MIT Press, 2018), 109-114.

[78] Goldstein, Joseph. *Mindfulness: A Practical Guide to Awakening* (Sounds True, 2016), 60.

[79] Tou, Hseuh. *The Blue Cliff Record*. Translated by Thomas Cleary & J.C. Cleary (Shambhala, 2005), 259.

[80] Goldstein, Joseph. *Mindfulness: A Practical Guide to Awakening* (Sounds True, 2016), 112.

[81] Tou, Hseuh. *The Blue Cliff Record*. Translated by Thomas Cleary & J.C. Cleary (Shambhala, 2005), 411.

[82] "Hubble Reveals Observable Universe Contains 10 Times More Galaxies Than Previously Thought," NASA, accessed September 20, 2022. https://www.nasa.gov/feature/goddard/2016/hubble-reveals-observable-universe-contains-10-times-more-galaxies-than-previously-thought.

[83] Hawking, Stephen, and George Ellis. *The Large Scale Structure of Space-Time* (Cambridge University Press, 1973), 364.

[84] Ibid., 128.

[85] Collins, Christopher, and Stephen Hawking. "*Why Is the Universe Isotropic?*" Astrophysical Journal, 180 (1973): 317–34.

[86] Hawking, Stephen, and George Ellis. *The Large Scale Structure of Space-Time* (Cambridge University Press, 1973), 360; Carroll, Sean. *From Eternity to Here: The Quest for the Ultimate Theory of Time* (Dutton, 2010), 351.

[87] Goldstein, Joseph. *Mindfulness: A Practical Guide to Awakening* (Sounds True, 2016), 133.

[88] Ryan, Sean, and Andrew Norton. *Stellar Evolution and Nucleosynthesis* (Cambridge University Press, 2010), 194.

[89] Thorne, Kip. *Black Holes and Time Warps: Einstein's Outrageous Legacy* (Norton, 1994), 168.

[90] Ibid., 206.

[91] NASA, "The Pillars of Creation," NASA.gov, accessed May 17, 2023, https://www.nasa.gov/image-feature/the-pillars-of-creation; NASA, "Helix Nebula: Blanco and Hubble," NASA Science, accessed May 17, 2023, https://science.nasa.gov/helix-nebula-blanco-and-hubble; NASA, "The Tycho Supernova: Death of a Star," NASA.gov, accessed May 17, 2023, https://www.nasa.gov/image-feature/the-tycho-supernova-death-of-a-star.

[92] Goldstein, Joseph. *Mindfulness: A Practical Guide to Awakening* (Sounds True, 2016), 126.

[93] Tou, Hseuh. *The Blue Cliff Record.* Translated by Thomas Cleary & J.C. Cleary (Shambhala, 2005), xiii.

[94] Rubin, Rick. *The Creative Act: A Way of Being* (Penguin Press, 2023), 15.

[95] "Nucleosynthesis periodic table," Wikipedia Commons, accessed May 17, 2023. https://commons.wikimedia.org/w/index.php?curid=31761437

[96] Thorne, Kip. *Black Holes and Time Warps: Einstein's Outrageous Legacy* (Norton, 1994), 450.

[97] Hawking, Stephen, and George Ellis. *The Large Scale Structure of Space-Time* (Cambridge University Press, 1973), 300

[98] Ibid., 321; Thorne, Kip. *Black Holes and Time Warps: Einstein's Outrageous Legacy* (Norton, 1994), 414.

[99] Ibid., 293.

[100] Ibid., 246.

[101] Ibid., 376.

[102] Ibid., 487, 489; Hawking, Stephen, and George Ellis. *The Large Scale Structure of Space-Time* (Cambridge University Press, 1973), 360.

[103] Goldstein, Joseph. *Mindfulness: A Practical Guide to Awakening* (Sounds True, 2016), 65.

[104] "Planck: Cosmic Recipe," *European Space Agency* (ESA). Accessed July 11, 2022 https://www.esa.int/ESA_Multimedia/Images/2013/03/Planck_cosmic_recipe..

[105] Mo, Houjun, Frank van den Bosch, and Simon White. *Galaxy Formation and Evolution* (Cambridge University Press, 2010), 7-13.

[106] Ibid., 640.

[107] Thorne, Kip. *Black Holes and Time Warps: Einstein's Outrageous Legacy* (Norton, 1994), 337..

[108] Ibid., 345.

[109] Konchady, Tarini. "Seeing Star Formation at Cosmic Noon," AAS Nova, May 14, 2021, https://aasnova.org/2021/05/14/seeing-star-formation-at-cosmic-noon; Nielsen, Nikole et al., "The CGM at Cosmic Noon with KCWI: Outflows from a Star-forming Galaxy at z = 2.071," The Astrophysical Journal, 904:164 (22pp), 2020 December 1. https://iopscience.iop.org/article/10.3847/1538-4357/abc561/pdf

[110] Goldstein, Joseph. *Mindfulness: A Practical Guide to Awakening* (Sounds True, 2016), 158.

[111] Schröder, K.P., "Distant future of the Sun and Earth revisited," arXiv preprint, arXiv:0801.4031, Jan 20, 2008, https://arxiv.org/pdf/0801.4031.pdf

[112] Mack, Katie. *The End of Everything* (Astrophysically Speaking) (Scribner, 2020), 51-70.

[113] Ibid., 71-104.

[114] Ibid., 105-128.

[115] Sekida, Katsuki. *Zen Training* (Shambhala Classics, 1985), 47.

[116] Harari, Yuval. *Sapiens: A Brief History of Humankind* (HarperCollins, 2015), 8.

[117] Taleb, Nassim. *Skin in the Game: Hidden Asymmetries in Daily Life* (Random House, 2018), 149.

[118] Ibid., 143, 149.

[119] Pickover, Clifford. *The Math Book: From Pythagoras to the 57th Dimension, 250 Milestones in the History of Mathematics* (Sterling, 2009), 18.

[120] Popper, Karl. *The Open Society and Its Enemies* (Princeton University Press, 1945), 230.

[121] IndoEuropeanTree," Wikipedia Commons, accessed May 17, 2023. https://en.wikipedia.org/wiki/Historical_linguistics#/media/File:IndoEuropeanTree.svg

[122] Dawkins, Richard. *The Selfish Gene* (Oxford University Press, 1976), 235-236.

[123] Nick Szabo, "Shelling Out: The Origins of Money," Nakamoto Institute, September 21, 2002, https://nakamotoinstitute.org/shelling-out/.

[124] Harari, Yuval. *Sapiens: A Brief History of Humankind* (HarperCollins, 2015), 115.

[125] Watts, Alan. *The Way of Zen* (Vintage Books, 1957), 142.

[126] "Animal Social Behaviour," in Encyclopedia Britannica, accessed May 17, 2023, https://www.britannica.com/topic/animal-social-behaviour.

[127] Harari, Yuval. *Sapiens: A Brief History of Humankind* (HarperCollins, 2015), 102-110.

[128] "Code of Hammurabi," in Encyclopedia Britannica, accessed May 17, 2023, https://www.britannica.com/topic/Code-of-Hammurabi.

[129] Goldstein, Joseph. *Mindfulness: A Practical Guide to Awakening* (Sounds True, 2016), 121.

[130] Tou, Hseuh. *The Blue Cliff Record*. Translated by Thomas Cleary & J.C. Cleary (Shambhala, 2005), 211.

[131] Watts, Alan. *The Way of Zen* (Vintage Books, 1957), 101-102.

[132] Tou, Hseuh. *The Blue Cliff Record*. Translated by Thomas Cleary & J.C. Cleary (Shambhala, 2005), 157.

[133] Tou, Hseuh. *The Blue Cliff Record*. Translated by Thomas Cleary & J.C. Cleary (Shambhala, 2005), 350, 462, 484, 488.

[134] Diamond, Jared. *Guns, Germs, and Steel: The Fates of Human Societies*. Norton, 1997.

[135] Popper, Karl. *The Open Society and Its Enemies* (Princeton University Press, 1945), 109.

[136] Ibid., 149.

[137] Whyte, David. *Consolations: The Solace, Nourishment and Underlying Meaning of Everyday Words* (Many Rivers Press, 2019), 214.

[138] Popper, Karl. *The Open Society and Its Enemies* (Princeton University Press, 1945), 173.

[139] Davidson, James, and William Rees-Mogg. *The Sovereign Individual: Mastering the Transition to the Information Age* (Simon and Schuster, 1999), 145.

[140] Ibid., 225.

[141] Ridley, Matt. *The Rational Optimist: How Prosperity Evolves* (HarperCollins, 2010), 182.

[142] Davidson, James, and William Rees-Mogg. *The Sovereign Individual: Mastering the Transition to the Information Age* (Simon and Schuster, 1999), 233.

[143] Ibid., 228.

[144] Ibid., 226.

[145] Ibid., 234.

[146] Ibid., 209.

[147] Ibid., 267.
[148] Ibid., 200.
[149] Ibid., 219.
[150] Ibid., 209.
[151] Ridley, Matt. *The Rational Optimist: How Prosperity Evolves* (HarperCollins, 2010), 182.
[152] Popper, Karl. *The Open Society and Its Enemies* (Princeton University Press, 1945), 230.
[153] Hawking, Stephen, and George Ellis. *The Large Scale Structure of Space-Time* (Cambridge University Press, 1973), 360; Carroll, Sean. *From Eternity to Here: The Quest for the Ultimate Theory of Time* (Dutton, 2010), 229.
[154] Griffiths, David, and Darrell Schroeter. *An Introduction to Quantum Mechanics*, Third Edition, (Cambridge University Press, 2018), 463.
[155] Thorne, Kip. *Black Holes and Time Warps: Einstein's Outrageous Legacy* (Norton, 1994), 95.
[156] Popper, Karl. *The Open Society and Its Enemies* (Princeton University Press, 1945), 61.
[157] Richard P. Feynman, Robert B. Leighton, and Matthew Sands, *Probability*, The Feynman Lectures on Physics, Volume 1, Online edition, Caltech, accessed May 17, 2021 https://www.feynmanlectures.caltech.edu/I_12.html..
[158] Richard P. Feynman, Robert B. Leighton, and Matthew Sands, *Characteristics of Force*, The Feynman Lectures on Physics, Volume 1, Online edition, Caltech, accessed May 17, 2021. https://www.feynmanlectures.caltech.edu/I_12.html.
[159] Richard P. Feynman, Robert B. Leighton, and Matthew Sands, *Characteristics of Force*, The Feynman Lectures on Physics, Volume 1, Online edition, Caltech, accessed May 17, 2021. https://www.feynmanlectures.caltech.edu/I_12.html.
[160] Popper, Karl. *Conjectures and Refutations* (Routledge & Kegan Paul, 1963), 9.
[161] Watts, Alan. *The Way of Zen* (Vintage Books, 1957), 154.
[162] Goldstein, Joseph. *Mindfulness: A Practical Guide to Awakening* (Sounds True, 2016), 323.

INDEX

Accretion disk 201, 202, 220, 221, 285
Anattā 244, 285
Anicca 244, 285
Antimatter 285
Arahant 100, 285
Atom 38, 53, 73, 108, 112, 140, 178, 179, 181, 200, 228, 258, 285, 287, 289, 290
Awakening 33, 34, 44, 48, 49 50, 299
Baryons 285
Beginner's mind 21, 39, 40, 285
Bhikkhus 204, 244-247, 285
Big Bang 89, 122, 126, 134-136, 141, 143, 285
Big Bang model 198, 285
Black Hole 14, 38, 42, 80, 126, 140, 163, 187, 193, 194, 201-203, 208-215, 219-222, 264, 285, 288, 291
Bosons 179, 181, 285, 288
Brainblast 203, 205, 206, 286, 287
Brown dwarf 201, 202, 286
Buddha 43, 44, 49, 51, 62, 63, 81, 104, 109, 117, 177, 203, 217, 244, 246, 248, 264, 266, 286, 287
Cantor's Diagonal Argument 74, 76, 78, 95, 98, 286

Cantor Set 89, 91-99, 110, 113, 114, 158, 286, 292
Cardinality 72, 74, 286
Choice point 187, 286
Clopen 6, 16, 53, 56, 57, 60-64, 82, 87, 101, 102, 120, 125, 149, 150, 157, 168, 17-178, 182, 186-191, 218, 227, 235, 257, 264, 277, 286, 287, 289
Closed idea space 53-56, 61, 62, 86, 87, 286
Closed society 251, 257, 286
Complement 12, 57, 58, 60, 64, 101, 286, 287
Consciousness 12, 14, 16, 17, 23, 38, 40, 41, 50, 57, 107, 109, 118, 124, 172, 240, 244, 286, 287
Cosmic Microwave Background (CMB) 136, 137, 198, 286
Countable 73, 74, 76, 94, 95, 98, 114, 286, 292
Dark energy 79, 138-140, 142, 182, 219, 228, 286
Dark matter 182, 219, 220, 224, 286
Devil's Staircase 157-159, 164, 287
Dharma 244, 248, 264, 287
Dukkha 244-247, 287

Einstein Equivalence Principle 161, 287
Electric charge 287
Electron 108, 178, 180, 181, 185, 186, 258, 285, 287
Elementary ideas 37, 38, 40, 41, 52, 81, 99, 193, 287
Empty set, ∅ 35, 37, 39, 60, 61, 287
Enlightenment 12, 15, 34, 44, 104, 204-207, 222, 224, 244, 264, 285-287, 289
Euclidean geometry 152-154, 287, 290
Event 42, 128-131, 135, 163, 167, 187, 195, 198, 199, 204, 208-210, 220, 268, 286, 287, 289
Everything 14-16, 26, 27, 36-38, 40, 51, 53, 54, 60, 64, 67, 77, 81, 82, 85, 89, 90, 95, 101, 102, 103, 109, 120, 121, 125, 126, 128-135, 142, 144, 146, 147, 151, 152, 157, 159, 164, 176-178, 182, 187, 188, 193, 196, 198, 207, 217, 224, 227, 228, 230, 255, 261, 266, 268, 271, 275, 282, 287, 290, 292, 297, 300, 301, 305
Fermions 179, 180, 287
Fission 287
Fractals 110-112, 123, 134, 142, 152, 196, 287, 290
Free will 187, 287
Function 71, 72, 74, 158, 241, 285, 288, 292
Fundamental ideas 41, 43, 81, 287
Fusion 201, 287, 289

General Relativity 14, 68, 74, 78, 151, 152, 162, 164, 195, 213, 260, 288, 290, 291, 296, 302
Gluon 180, 181, 288
Grand Unification Theory 288
Gravitational wave 213, 288
Graviton 143, 181, 288
Hadrons 180, 288
Headless Way 87, 107, 110, 118, 121, 123, 124, 146, 288
Higgs boson 181, 288
Horizon 42, 125, 147, 195, 196, 208-210, 220, 288
Hubble Radius 134, 140-142, 228, 288
Idea space 11, 12, 14-18, 22, 23, 25-31, 34-41, 43-51, 53-56, 61-64, 68, 69, 79-81, 85-89, 91, 98-100, 102-105, 107, 108, 110, 117, 120, 122, 126, 133, 143, 144, 146, 147, 149, 151, 154, 159, 168, 171, 172, 176-178, 187-190, 193-195, 198-200, 202, 206-208, 214-219, 222-224, 226-233, 235, 236, 239, 241, 244-246, 248, 249, 251-254, 256, 258, 259, 261, 262, 264, 266, 267, 270-276, 278, 282, 285-292
Identity 14, 34, 48, 49, 103, 107-109, 112, 113, 120, 123, 217, 229, 264, 266, 288, 289
Illusion of Self 16, 107, 122, 288
Impermanence 16, 49, 67, 68, 79, 81, 82, 193, 218, 229, 230, 244, 246, 285, 288
Inflation 288

INDEX

Information Age 256, 257, 271, 273, 288, 295, 306
Injection 288
Insights 12, 124, 152, 203, 205, 288
Intersubjective realities 16, 17, 240, 256, 288
Known Knowns 41, 42, 287, 289
Known Unknowns 41, 42, 287, 289
Koan 48-50, 89, 102-105, 107, 170, 203, 205, 206, 208, 214, 217, 218, 222-224, 226, 227, 229, 264, 276, 278, 287, 288, 290
Light cone 128-133, 141, 195, 289
Lindy Test 232, 233, 235, 236, 238-240, 243, 248, 249, 256, 289
Macro idea space 28, 29, 45, 146, 231-233, 241, 251-254, 256, 272, 273, 288
Main sequence 201-203, 286, 289
Meditation 12, 14, 38, 85, 86, 87, 101, 107, 117, 124, 150, 188, 189, 204, 205, 231, 289, 293, 300
Meson 289
Mettā 87, 247, 281, 289
Mindfulness 12, 13, 15, 16, 51, 52, 61, 64, 85-89, 100-103, 110, 117, 124, 148-150, 172, 175, 187-191, 200, 217, 218, 226, 227, 247, 248, 259, 261, 288, 289, 291, 293, 299, 300, 302-307
Nāmarūpa 57, 149, 289
Nebula 201, 289, 303
nen 175

Neutrinos 180, 181, 288, 289, 296, 302
Neutron 14, 180, 181, 201-203, 286, 289
Neutron star 201-203, 286, 289
Nirvana 248, 264, 289
Nonduality 16, 51, 53, 57, 63, 101, 176-178, 289
Non-Euclidean geometry 152, 153, 154, 290
Non-Self 4, 9, 10, 14-17, 34, 44, 107-110, 122, 123, 134, 142, 143, 146, 196, 217, 227, 245, 266, 277, 278, 288, 290
Nucleus 178, 180, 181, 258, 285, 289, 290
Observable universe 14, 16, 35, 36, 50, 89, 108, 112, 122, 126, 133-136, 140-143, 147, 188, 189, 193-198, 219, 271, 272, 285-292
One-Nen 175, 290
Open idea space 53, 56, 61, 62, 87, 252, 290
Open society 251, 257, 290
Particle-wave duality 16, 64, 178, 182, 186-189, 290
Personal Idea Space 28, 44, 45, 231, 290
Photon 181, 290
Practice 12, 14, 15, 51, 85-87, 101, 117, 148, 149, 172, 188-190, 204, 205, 264, 289, 290
Principle 104, 136, 157, 161, 164, 186, 210, 218, 222-224, 226, 227, 231, 264, 287, 290-292
Quantum mechanics 38, 64, 186, 258, 290

Quarks 112, 180, 181, 187, 219, 285, 287-290
Recombination 136, 198, 285, 290
Reference frame 77, 159, 165, 167, 168, 291
Samādhi 291
Schwarzschild Radius 208, 209, 211, 291
Science 4, 9-13, 17, 52, 75, 117, 118, 120-125, 134, 146, 147, 202, 230, 236, 248, 249, 258, 261, 262, 291, 295, 296, 302, 303
Science of objects 10, 11, 12, 117, 118, 120-123, 125, 134, 146, 147, 202, 230, 248, 261, 262, 291
Science of the first person 11, 12, 117, 118, 120-125, 134, 146, 147, 202, 230, 248, 249, 261, 262, 291
Set 13, 16, 17, 20, 23, 35-40, 50, 51, 56-61, 70, 72-74, 88, 89, 91-98, 110, 113, 114, 117, 125, 135, 147, 158, 182, 227, 235, 242, 254, 286-288, 291, 292
Singularity 16, 88, 89, 90, 91, 98, 99, 100, 103, 104, 108, 126, 135, 143, 144, 146, 148, 169, 177, 195-198, 208, 210, 212, 213, 218, 224, 227, 228, 236, 272, 275, 290-292
Singularity Sunset 108, 126, 135, 143, 146, 148, 169, 196, 272, 291, 292

Source 63, 82, 135, 136, 137, 143, 149, 185, 186, 196, 227, 239, 272, 283, 291
Spacetime 10, 16, 50, 69, 74, 78, 79, 104, 128, 129, 145-147, 150, 151, 152, 157-159, 161-165, 168-170, 172, 177, 181, 194, 195, 197, 201, 206-209, 211, 213, 214, 215, 217-219, 221, 227, 243, 264, 272, 275-277, 285, 287, 288, 291, 292, 296, 297, 302
Special relativity 151, 157, 161, 164, 287, 291
Standard model of physics 291
Stoicism 87, 117, 203, 291
Sunset Conjecture 16, 108, 125, 126, 143, 144, 146, 147, 154, 292
Supernova 181, 201, 203, 205, 222, 287, 292, 303
Theory of 292, 296, 298, 300-303, 307
Topological 16, 88-91, 98, 99, 100, 102-104, 110, 143, 144, 171, 177, 196-198, 218, 224, 228, 236, 248, 275, 277, 289, 292
Uncertainty principle 186, 290, 292
Uncountable 14, 16, 25, 27, 35, 44, 50, 68, 69, 73-82, 88, 89, 90, 91, 94, 95, 98, 99, 100, 102, 103, 112, 114-116, 120, 126, 143, 147, 149, 193, 197, 235, 246, 275, 286, 288, 292

Universe 9, 10, 12, 14, 16, 34-36, 50, 67, 68, 73, 78, 79, 89, 103, 108, 112, 117, 122, 123, 126, 133-143, 145-147, 149, 152, 155, 159, 172, 178, 179, 181, 182, 187-189, 193-198, 200, 206, 207, 209, 210, 213, 219-222, 224, 227, 228, 230, 236, 246, 258, 259, 260, 262, 271, 272, 275, 285, 286, 288-292, 296, 297, 301, 302, 303

Unknown Knowns 41, 42, 287, 292

Unknown Unknowns 15, 17, 41, 42, 43, 171, 264, 287, 292

Veils of Illusion 15, 17, 44, 50, 171, 172, 266, 277, 292

$W^{\pm}$ and Z^0 Boson 292

Weak Equivalence Principle 161, 292

White dwarf 201, 202, 209, 288, 292

Worldline 129, 169, 170, 292

Wormhole 213, 214, 216-218, 229, 264, 268, 289, 292

Yathabhutam 39, 218, 247, 277, 292

Zero measure 11, 14, 16, 23, 25, 35, 36, 40, 44, 50, 79, 88-91, 94, 98, 99, 102, 103, 108, 114-117, 121, 122, 125, 126, 143,-145, 147, 177, 193, 196, 218, 230, 235, 261, 286, 288, 290-292

Ready to extend your mindfulness practice to new grounds?

Cards

Increase relaxation and amplify personal insights with **100 Mindful Prompts** and **100 Daily Meditations**.

100 Mindful Prompts   100 Daily Meditations

Build a Mindful Base
with the Idea Space

Shirts

Make a statement with our comfortable and stylish shirts designed to inspire and motivate you every day.

ABOUT THE AUTHOR

"Life is but a series of paths."
- Clément Decrop

Inventor and Belgium-born author, Clément Decrop moved to the United States at the age of six with his family. With a degree in Mechanical Engineering from Penn State, Decrop has worked across the globe, including France, Spain, the United Emirates, and then back home in the United States.

The Idea Space: The Science of Awakening Your Non-Self delves into the depths of consciousness by introducing a distinctive solution to Einstein's field equation to describe the mind, accessible to the layperson. In these pages, Decrop guides readers to view their thoughts objectively and identify their impact, helping them discover a happier existence and a deeper understanding of their life's purpose.

As a *Global Educator* since 2018, Clément has shared his wisdom on meditation, sleep, exercise, and nutrition with thousands of eager participants across 40 countries. His innovative spirit led him to collaborate with numerous inventors from Wikipedia's Most Prolific Inventors List, resulting in 130-plus patent disclosures within one year, 50-plus filed, and 25-plus issued as of late 2023.

Outside these professional spheres, Decrop finds joy in the simpler pleasures of life. He's an avid reader—from textbooks on semiconductors and rockets to general fiction—an experimental cook, a global traveler, an enthusiastic coder, and a health & fitness devotee. He lives in Pennsylvania.

Made in United States
Troutdale, OR
11/25/2023

14921357R00195